T0344589

Advances in Structures, Properties and Applications of Biological and Bioinspired Materials

MATERIALS RESEARCH SOCIETY
SYMPOSIUM PROCEEDINGS VOLUME 1621

Advances in Structures, Properties and Applications of Biological and Bioinspired Materials

Symposia held December 1–6, 2013 Boston, Massachusetts. U.S.A.

EDITORS

Syam P. Nukavarapu
University of Connecticut
Farmington, Connecticut, U.S.A

Huinan Liu
University of California, Riverside
Riverside, California, U.S.A.

Tao Deng
Shanghai Jiao Tong University
Shanghai, China

Michelle Oyen
Cambridge University
Cambridge, United Kingdom

Candan Tamerler
University of Kansas
Lawrence, Kansas, U.S.A.

Materials Research Society
Warrendale, Pennsylvania

CAMBRIDGE
UNIVERSITY PRESS

Shaftesbury Road, Cambridge CB2 8EA, United Kingdom

One Liberty Plaza, 20th Floor, New York, NY 10006, USA

477 Williamstown Road, Port Melbourne, VIC 3207, Australia

314–321, 3rd Floor, Plot 3, Splendor Forum, Jasola District Centre, New Delhi – 110025, India

103 Penang Road, #05–06/07, Visioncrest Commercial, Singapore 238467

Cambridge University Press is part of Cambridge University Press & Assessment, a department of the University of Cambridge.

We share the University's mission to contribute to society through the pursuit of education, learning and research at the highest international levels of excellence.

www.cambridge.org
Information on this title: www.cambridge.org/9781605115986

Materials Research Society
506 Keystone Drive, Warrendale, PA 15086
http://www.mrs.org

First published 2014

A catalogue record for this publication is available from the British Library

CODEN: MRSPDH

ISBN   978-1-605-11598-6   Hardback

CONTENTS

ADVANCES IN MECHANICS OF BIOLOGICAL AND BIOINSPIRED MATERIALS

ENGINEERING AND APPLICATION OF BIOINSPIRED STRUCTURED MATERIALS

MATERIALS FOR NEURAL INTERFACES

*Invited Paper

PREFACE

There were eleven symposia conducted with a focus on biomaterials under the sub-class of "Biomaterials and Soft Matter" in the MRS 2013 fall meeting. Symposia, H, C, D and J, with a considerable overlap in the content and focus have been combined into single proceedings entitled "Advances in Structures, Properties and Applications of Biological and Bioinspired Materials". This volume contains both invited as well as regular submissions to the symposia listed below:

Symposium H, "Advanced Composites and Structures for Tissue Engineering"
Symposium C, "Advances in Mechanics of Biological and Bioinspired Materials"
Symposium D, "Engineering and Application of Bioinspired Structured Materials"
Symposium J, "Materials for Neural Interfaces"

The aim of this volume is to provide state-of-the-art research in biomaterials and bio-inspired materials—their structures, properties, and applications. Biological materials provide the critical support for the proper function of the living systems. The wide range and complexity exhibited by biological materials is unmatched in current bio-inspired as well as biomaterials. Understanding the underlying operation principles of these materials will provide promising approaches to not only improve biomaterials and bio-inspired materials' performances in the living systems, but also provide creative new materials and systems that enable the applications of those materials for bio-sensing to tissue engineering. This symposia Proceedings volume presents some of the most recent advancements in the following areas: (1) Tissue Engineering, (2) Bio-inspired Materials, (3) Bio-inspired structures, and (4) Neural Interfaces. We hope the papers in the volume will provide the readers some insight of the current methodologies and approaches used in studying fundamental problems associated with biological materials, biomaterials, and bio-inspired materials. Some stimulating discussions of potential applications of these materials are also offered in the volume, which represents the trend of research in this exciting area of materials science and engineering.

Syam Nukavarapu
Huinan Liu
Tao Deng
Michelle Oyen
Candan Tamerler

May 2014

MATERIALS RESEARCH SOCIETY SYMPOSIUM PROCEEDINGS

MATERIALS RESEARCH SOCIETY SYMPOSIUM PROCEEDINGS

MATERIALS RESEARCH SOCIETY SYMPOSIUM PROCEEDINGS

Volume 1663E – Self-Organization and Nanoscale Pattern Formation, 2014, M.P. Brenner, P. Bellon, F. Frost, S. Glotzer, ISBN 978-1-60511-640-2

Volume 1664E – Elastic Strain Engineering for Unprecedented Materials Properties, 2014, J. Li, E. Ma, Z.W. Shan, O.L. Warren, ISBN 978-1-60511-641-9

Prior Materials Research Symposium Proceedings available by contacting Materials Research Society

Advanced Composites and Structures for Tissue Engineering

Mater. Res. Soc. Symp. Proc. Vol. 1621 © 2014 Materials Research Society
DOI: 10.1557/opl.2014.4

Composites and Structures for Regenerative Engineering

Cato T. Laurencin[1,2,3,4,*], M.D., Ph.D. and Roshan James[1,2,4], Ph.D.

[1]Institute for Regenerative Engineering
[2]The Raymond and Beverly Sackler Center for Biomedical, Biological, Physical and Engineering Sciences
[3]Connecticut Institute for Clinical and Translational Science
[4]Department of Orthopaedic Surgery

The University of Connecticut Health Center
263 Farmington Avenue
Farmington, CT 06030, U.S.A

*Corresponding author:
Cato T. Laurencin, M.D.,Ph.D.
University Professor
University of Connecticut Health Center
263 Farmington Avenue, Farmington, CT, USA 06030
Phone: +1-860-679-6544
Fax: +1-860-679-1553
Email: laurencin@uchc.edu

ABSTRACT

Regenerative engineering was conceptualized by bridging the lessons learned in developmental biology and stem cell science with biomaterial constructs and engineering principles to ultimately generate de novo tissue. We seek to incorporate our understanding of natural tissue development to design tissue-inducing biomaterials, structures and composites than can stimulate the regeneration of complex tissues, organs, and organ systems through location-specific topographies and physico-chemical cues incorporated into a continuous phase. This combination of classical top-down tissue engineering approach with bottom-up strategies used in regenerative biology represents a new multidisciplinary paradigm. Advanced surface topographies and material scales are used to control cell fate and the consequent regenerative capacity.

Musculoskeletal tissues are critical to the normal functioning of an individual and following damage or degeneration they show extremely limited endogenous regenerative capacity. The increasing demand for biologically compatible donor tissue and organ transplants far outstrips the availability leading to an acute shortage. We have developed several biomimetic structures using various biomaterial platforms to combine optimal mechanical properties, porosity, bioactivity, and functionality to effect repair and regeneration of hard tissues such as bone, and soft tissues such as ligament and tendon. Starting with simple structures, we have developed composite and multi-scale systems that very closely mimic the native tissue architecture and material composition. Ultimately, we aim to modulate the regenerative potential, including proliferation, phenotype maturation, matrix production, and apoptosis through cell-scaffold and host –scaffold interactions developing complex tissues and organ systems.

INTRODUCTION

Health care issues arising from tissue loss or organ failure are among the most devastating and costliest world over [1]. Musculoskeletal injuries are largely from trauma, disease, and birth defects, and collectively there occurs over 34 million musculoskeletal organ repair or replacement surgeries in the US each year. The fascination to regenerate tissues moved toward a realistic platform when Y. C. Fung first proposed the term 'Tissue Engineering' at the 1987 meeting of the National Science Foundation [2]. The state of art today has been the use of autografts and/or allografts, and currently we are poised to move forward to the next level especially where we need to address the repair of severely damaged limbs [3]. 'Tissue Engineering' is defined as the use of isolated cells or cell substitutes, tissue-inducing substances, and cells placed on or in matrices to repair and regenerate tissue [4]. These pioneering efforts led to the realization of regenerative medicine that was only previously depicted in literature, art, science fiction and in comic books.

Laurencin defined 'Regenerative Engineering' as taking tissue engineering a step further by integrating it with advanced materials science, stem cell science, and areas of developmental biology [5, 6]. The grand challenge of the future is to achieve complete limb regeneration, for example, in a newt an injured or chopped limb will be completely regenerated within a period of 10 weeks [7]. Whereas tissue engineering involves interdisciplinary teams from the fields of engineering, science, and medicine, we see regenerative engineering as an expansion of this approach—a "convergence" [8] of tissue engineering with nanotechnology, stem cells, and developmental biology [7]. This new frontier, enabling the reconstruction of complex tissues and whole organs, requires the combination of top-down engineering approaches with bottom-up strategies that integrate materials science and tissue engineering with stem cell and developmental biology [5]. The emergence of regenerative strategies represents both technical advances as well as an evolution of patient preferences. Conventional treatments such as total knee replacement, total shoulder replacement, urostomy bag, and pacemaker may be replaced by innovative alternatives such as restorable meniscal matrices, regenerative rotator cuff scaffolds, functional bladder, and pacemaker cells respectively.

The limb is a complex organ composed of complicated structures including skin, bone, cartilage, muscle, nerves, and blood vessels. The challenge is to regenerate multiple tissue types simultaneously into one intact and functioning organ. The current dogma in developmental biology is that regenerative capacity is inversely related to organismal complexity. A key element is to utilize biomaterials as guides for tissue development, where we will engineer substrates that through topographical and physicochemical means will influence the differentiation of undifferentiated cell types. It seems clear that defining and controlling the micro-environmental niche within each tissue and even within regions of a given tissue is key to successful tissue regeneration. The niche conditions include factors such as oxygen concentration, cytokine gradients, pH, ionic and electrical potential, available nutrients, and mechanical forces, all of which are maintained in a state of dynamic equilibrium in temporal and spatial patterns unique to each tissue and organ. A 3-dimensional (3D) platform while more complex, better mimics the *in vivo* organization of cells and is a crucial requirement for tissue construction [9]. The lack of guidance cues in monolayer culture techniques can be overcome using advanced biomaterials, new designs, materials with tunable physical properties such as stiffness, and by incorporating chemical modifications to include stimulatory factors. Most niches are a 3D network within which the cells are fully embedded, and both synthetic and natural materials are used to mimic the 3D native architecture. This manuscript presents our advances in the regeneration of diverse musculoskeletal tissues laying the foundations of multi-tissue and organ regeneration.

ADVANCED BIOMATERIALS AND STRUCTURES

The components required for inducing regeneration are an appropriate biomaterial, instructive structures, cells, and soluble regulator factors. These cues have been investigated in various permutations and combinations to induce repair and regeneration of various tissues, and skin, nerves and bone are the most commonly studied tissues. In one approach, a donor organ is decellularized and the extracellular matrix (ECM) of the tissue is used as a scaffold which is then seeded with patients own cells, e.g. bone marrow stromal cells (BMSCs) to create a functional substitute [10, 11]. This approach using one's own cells mitigates the immune response that occurs during conventional transplantation, however the availability of organs that can be used for decellularization continues to be restrictive [12]. These limitations have inspired scientists to develop natural and synthetic polymer derived scaffolds onto which cells can be cultured, and guided to specific phenotypes [13, 14]. The recent advances in biomaterial synthesis and fabrication tools have made it possible to guide cells into complex, three-dimensional (3D) structures by using appropriate scaffolds templates. The increased understanding of the cell-substrate communications have driven efforts to develop scaffold systems that recapitulate key features of the ECM that control the migration, proliferation, and differentiation of cells.

The ECM is composed of interwoven fibers made mostly of fibrillar collagen and elastin, with diameters ranging from tens to hundreds of nanometers. The ECM sequesters and presents cytokines and specific binding sites for cellular adhesion [15]. Cross communication between cells and the ECM occurring through interactions between ligands and cell surface receptors define the microenvironment that control behavior and fate of the cell [7, 16]. Integrating topographical cues is especially important in engineering complex tissues that have multiple cell types and require precisely defined cell-cell and cell-matrix interactions in a 3D environment [16-18]. Thus, in a regenerative engineering approach, micro and nanoscale materials/structures play a paramount role in controlling cell fate and the consequent regenerative capacity.

HARD MUSCULOSKELETAL TISSUE

The hierarchical organization of bone tissue comprising of woven and lamellar bone, interstitial networks and gap-junctions is very difficult to exactly mimic when developing bone substitutes to modulate repair/regeneration. Laurencin *et. al.* reported a bone tissue engineering scaffold which utilized heat sintering technique to fabricate sintered microsphere matrices having sufficient porosity for cellular migration and movement of fluids across the scaffolds. Using polymeric microspheres made of poly(lactide-*co*-glycolide) (PLAGA), 3D heat sintered scaffold having an interconnected pore structure that resembled the structure of trabecular bone was developed (Figure 1A) [19, 20]. The pore structure is a negative template of trabecular bone in structure and volume, and the newly forming bone would occupy the pore structure while the microsphere matrix slowly degraded leaving voids that will form the pore structure of newly formed trabecular bone. These sintered microsphere scaffolds can be tailor made with pore diameter, pore volume, and mechanical properties within a given range. Microspheres of diameter 600–710 μm were sintered yielding an optimal, biomimetic structure with pore diameter in the range of 83–300 μm onto which human osteoblasts seeded were cultured *in vitro*. On this particular matrix, the cells adhered and proliferated throughout the pore system (Figure 1B). The cells maintained bone phenotype expression on the above mentioned scaffold as evidenced by osteocalcin staining suggesting its osteoconductive potency (Figure 1C).

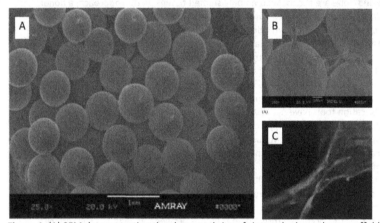

Figure 1. (A) SEM demonstrating the shape and size of sintered microsphere scaffold composed of diameter 600 – 710um diameter microspheres. (Magnification = 25X) [19]. "Reprinted from Clinical Orthopaedics & Related Research, 447, Cooper *et al.*, The ABJS Nicolas Andry Award: Tissue engineering of bone and ligament: a 15-year perspective, 221-236, 2006, with permission from Lippincott Williams & Wilkins" (B) Scanning electron micrographs showing the morphology of human osteoblasts at 16 days on the microsphere matrix. Micrograph demonstrates cellular adhesion within the matrix and the promotion of several cellular attachment sites between adjacent sintered microspheres [20]. (C) Immunofluorescence staining for osteocalcin to assess the osteoblast phenotypic behavior while cultured on the sintered matrix at 16 days. Micrograph demonstrates osteoblast cells with positive osteocalcin staining at various locations along the matrix [20]. "Reprinted from Biomaterials, 24 (4), Borden

et al., Structural and human cellular assessment of a novel microsphere-based tissue engineered scaffold for bone repair, 597-609, 2003, with permission from Elsevier"

Various grain sizes of hydroxyapatite (HA) which is similar to the inorganic component of bone have been combined with collagen and PLAGA to introduce osteoinductivity into the regenerative scaffold [21, 22]. Improved bone cell function has been correlated with reduced grain size of the HA constituent [16, 23, 24]. Compared to micro grain sized HA components, the particles of nano-dimensions further improved osteointegration with the host bone tissue. The native bone composition and structure is more closely mimicked by composites of nano HA (nHA) and collagen, and evidenced by the enhancement of bone cell function and host-integration. Stronger interfacial-bonding between the organic and inorganic phases was formed in composites having nHA resulting in higher mechanical properties as compared to those fabricated from micro or bulk-sized HA constituents [25]. These nanocomposites are successful in inducing better cellular responses as compared to standard composites due to their similarity with the natural bone structure, and additionally can be processed to have mechanical properties more closely to native bone. For example, Lv *et al.* have fabricated and optimized nHA-PLAGA composite scaffolds by varying several parameters such as polymer:ceramic ratio and sintering conditions [26]. A 20% nHA loading resulted in a significant increase in compressive mechanical properties which is desirable for bone regeneration applications. The incorporation of nHA contributed to the bioactivity of the scaffolds for better osteointegration. Under *in vitro* culture condition, human mesenchymal stem cells (hMSCs) expressed elevated expression of phenotypic markers such as alkaline phosphatase as well as increased mineral deposition on the nHA-PLAGA composite scaffolds. Integrating nanotopographical cues is important in engineering complex tissues that have multiple cell types and require precisely defined cell-cell and cell-matrix interactions in a 3D environment [18]. Thus, in a regenerative engineering approach, nanoscale materials/structures play a paramount role in controlling cell fate and the consequent regenerative capacity.

Nanotechnology has allowed for the fabrication of novel materials, structures, and systems with nanosize features for altered or increased biological responses compared with micrometer-size counterparts [27]. Various nanostructures have been fabricated including quantum dots and wires, nanoscale self-assemblies and thin films, nanocrystals, nanotubes, nanowires, nanorods, nanofoams, nanospheres and nanofibers [28]. Electrospun nanofiber matrices closely mimic the structure of natural ECM and show great promise for scaffold-based tissue engineering applications. Figure 2 shows a general representation of an electrospinning apparatus used to create polymeric nanofibers [27]. Laurencin et al. demonstrated the feasibility of developing nanofibrous structures from a variety of biodegradable polymers including poly(lactide-*co*-glycolide) (PLAGA), polyphosphazenes, and polymer composites [29-32]. The Journal of Biomedical Materials Research (JBMR) issued a special edition to commemorate the 100th volume (Progress in Biomaterials) featuring the top 25 papers of the last 50+ years and they recognized the pioneering paper by Laurencin *et. al.* which showed that biodegradable electrospun nanofiber structures were capable of supporting cell attachment and proliferation, and indicated that cells seeded on this structure could maintain phenotypic shape and guide growth according to nanofiber orientation [33].

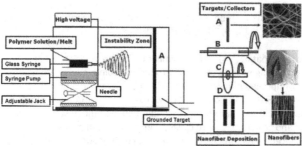

Figure 2. Schematics of electrospinning setup [27]: A high electric potential of a few kV is applied to the pendent polymer droplet/melts and a polymer jet is ejected from the charged polymer solution. The polymer jet undergoes a series of bending and stretching instabilities that causes large amounts of plastic stretching, resulting in ultra-thin fibers. Based on the type of collector/target and its motion, it is possible to align nanofibers. (A) Stationary collectors (grounded conductive substrate) result in random nanofibers. Rotating targets such as (B) a rotating drum and (C) wheel-like bobbin or metal frame result in fiber alignment and fabrication of tubular nanofiber scaffolds. (D) Example of a split electrode consists of two conductive substrates separated by a void gap that results in the deposition of aligned nanofibers across. "Reprinted from Biomedical Materials, 3 (3), Kumbar et al., Electrospun nanofiber scaffolds: engineering soft tissues, 034002, 2008, with permission from IOP Publishing"

Nanofiber scaffolds which are characterized by high porosity, high surface area, unusual surface properties, and morphological similarity to native bone ECM have been evaluated for bone tissue engineering applications. Various studies have documented the favorable responses of bone cells toward nanofiber or nanophase particles [29, 34-36]. Li *et al.* have demonstrated enhanced osteoblast adhesion and proliferation on polymeric nanofiber scaffolds that mimicked the ECM. The extracellular mimetic structure increased protein adsorption as well as to the selective adsorption of cell adhesion proteins such as fibronectin and vitronectin on the nanostructures compared to structures with solid pore walls [37]. Furthermore, the nanofiber scaffold environment has been shown to support mesenchymal stem cells to differentiate along the osteogenic lineage [38]. Laurencin *et al.* demonstrated that electrospun nanofiber matrices of poly[bis(*p*-methylphenoxy)phosphazene] (PNmPh) supported the adhesion and proliferation of osteoblast like MC3T3-E1 cells [30]. Bhattacharyya *et al.* fabricated poly[bis(ethyl alanato)phosphazene] (PNEA) as well as PNEA-hydroxyapatite composite nanofiber matrices as scaffolds for bone tissue engineering applications [31, 32]. Such polyphosphazene nanofiber structures closely mimic the ECM architecture, and have shown excellent osteoconductivity and osteointegration [30-32, 39].

Inspired by the hierarchical structures that enable bone function, Deng *et al.* recently fabricated a mechanically competent 3D scaffold mimicking the bone marrow cavity, as well as, the lamellar structure of bone by orienting electrospun dipeptide-based polyphosphazene-polyester blend nanofibers in a concentric manner with an open central cavity [40]. The 3D biomimetic scaffold exhibited a similar characteristic mechanical behavior to that of native bone. Compressive modulus of the scaffold was found to be within the range of human trabecular bone. As shown in Figure 3, *in vitro* studies using primary cell culture demonstrated the ability of the biomimetic scaffold to support the osteoblast proliferation and accelerated differentiation throughout the scaffold architecture, which resulted in a similar cell-matrix organization to that of native bone and maintenance of structure integrity. It was thus suggested that the concentric open macrostructures of nanofibers that structurally and mechanically mimic the native bone can be a potential scaffold design for accelerated bone healing.

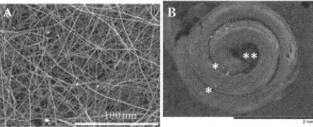

Figure 3. SEM images of (A) Biomimetic scaffolds comprised of electrospun nanofibers of dipeptide-based polyphosphazene-polyester blends mimicking the collagen fibrils present in native bone ECM. (B) The nanofiber mat rolled into a concentric circle and seeded with cells. ECM deposition is evident throughout 3D scaffold architecture during cell culture after 28 days of culture [40]. "Reprinted from Advanced Functional Materials, 21 (14), Deng *et al.*, Biomimetic Structures: Biological Implications of Dipeptide-Substituted Polyphosphazene–Polyester Blend Nanofiber Matrices for Load-Bearing Bone Regeneration, 2641–2651, 2011, with permission from John Wiley & Sons"

SOFT MUSCULOSKELETAL TISSUE

It is well established that soft musculoskeletal tissue injuries, such as involving the tendons and ligaments heal slowly and poorly because of limited vascularization, therefore requiring surgical intervention. Functional repair or regeneration of the rotator cuff, flexor tendons, and ACL remains extremely challenging. To overcome the limitations associated with current anterior cruciate ligament (ACL) grafts, such as lack of flexibility and tensile strength, a biodegradable ligament platform was designed based on the hierarchical complexity of natural ligament architecture [41]. Polymeric fibers measuring approximately 15 μm in diameter were bundled together to form yarns. These yarns were woven in circular or rectangular braids of varying yarn densities and braiding patterns (Figure 4) to achieve mechanical properties that best approximated a natural ligament. Several different shapes and yarn configurations were considered and evaluated through tensile testing to match the mechanical properties of the native tissue so as to function as efficiently as a native ligament. The results of tensile testing yielded stress/strain curves that mimicked the triphasic nature of the natural ligament. The synthetic scaffolds showed an initial low modulus followed by a linear region of increased modulus and ending in a plateau region, suggesting plastic deformation and ultimate failure. Synthetic ligaments also matched ultimate tensile strengths seen in native ligaments depending on the specific yarn density and braid shape. By varying these parameters it was possible to tune the ultimate tensile strength as needed.

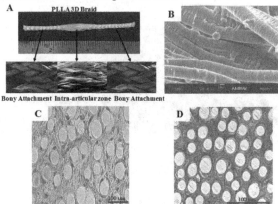

Figure 4. A hierarchical design for ligament regeneration [41]. (A) Design of 3D braided poly-L-lactide (PLLA) ligament scaffolds showing the macrostructure of the bony attachment and intra-articular zone. Briefly polymeric fibers measuring approximately 15 µm in diameter were bundled together to form yarns. These yarns were woven in circular or rectangular braids of varying yarn densities and braiding patterns to achieve mechanical properties that best approximated a natural ligament. The results of tensile testing yielded stress/strain curves that mimicked the triphasic nature of the natural ligament. (B) SEM micrograph showing guided rabbit ACL cell growth on PLLA scaffolds after 3 days of culture. *In vivo* assessment of the tissue-engineered ligament was done in male New Zealand White rabbits. (C) Unseeded synthetic ACL after 12 weeks of healing. Moderate collagen infiltration is evident between fiber bundles; and (D) Seeded synthetic ACL after 12 weeks of healing. Tissue infiltration was enhanced for the seeded synthetic ACL after 12 weeks of healing, suggesting over time the newly forming tissue migrated into the synthetic material and formed new tissue in the pore structure of the synthetic ligament. "Reprinted from Proceedings of the National Academy of Sciences of the United States of America, 104 (9), Cooper et al., Biomimetic tissue-engineered anterior cruciate ligament replacement, 3049–54, 2007, with permission from The National Academy of Sciences of the USA"

Through scaffold design based on structural mimicry of the native tissue ECM, Peach *et. al.* demonstrated that poly(ε-caprolactone) (PCL) fiber constructs of average fiber diameter 2900-3000nm offered the optimal balance of cell proliferation and tensile modulus over a period of 2 weeks in culture [42-44]. Furthermore, these optimal constructs were functionalized with 2% Poly[(ethyl alanato)$_1$(p-methyl phenoxy)$_1$] phosphazene (PNEA-mPh) a more hydrophilic polymer to increase total protein synthesis in human mesenchymal stem cells (hMSCs) cultured onto these scaffolds. Cell adhesion and long-term cell infiltration into the matrices were enhanced with the PNEA-mPh surface functionalization which is due to the introduction of a more roughened hydrophilic surface (Figure 5). Long-term tensile properties were unaffected by the functionalization, with cell seeding increasing initial tensile modulus and maintaining a modulus over 9 weeks culture within the range of clinically used materials. The PNEA-mPh functionalization led to overall greater tendon differentiation, with a greater expression of the mature tendon marker (Tenomodulin) and better tendon maturity as assessed by Collagen I:III expression ratio.

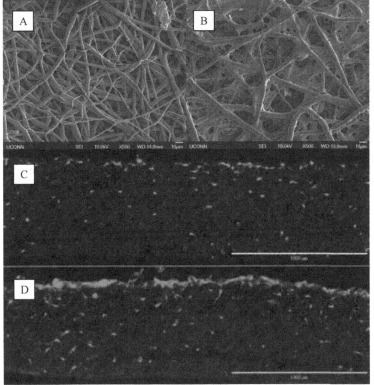

Figure 5. SEM imaging of PCL matrices functionalized with 0 and 2% PNEA-mPh solutions at 500X (A, B) magnifications [43]. Compared to the smooth morphology of the PCL control (A) the dipped scaffolds have what appear to be rough accumulations of PPHOS (B). Functionalization was achieved by dipping in 2%w/v PNEA-mPh in 50:50 AcO:EtOH for one second. Permeation of

hMSCs on non-functionalized PCL (C) and PCL electrospun fibers functionalized by a 2% solution of PNEA-mPh (D). Cells (green pseudo-color) were allowed to culture for 4 weeks to determine scaffold (blue pseudo-color) permeation. Functionalization does not inhibit cell infiltration from the scaffold surface. Scaffold seeded at 30,000 cells per 1×1 cm scaffold. Scale Bars = 1000 μm.

CONCLUSION

The composition of the biomaterial, the fabricated form, and the physico-chemical properties of the scaffold play an important role in the regenerative process. Scientists must develop an excellent understanding of cell behavior, and signaling mechanisms especially of cell-material interactions and how they modulate cellular events. The regeneration of complex tissues, organs or organ systems such as human limbs remains a significant challenge. The complexity within the native tissues and organs necessitates the development of biomimetic scaffold platforms that will elicit precise control of cellular events especially in regenerating various tissues in a continuous substrate. The focus needs to be on identifying the unique parameters that trigger specific signaling events, and improve understanding of the mechanistic pathways that drive development of tissues. Recapitulating the stem cell microenvironment using 3D scaffolds will maintain the stem cell population and guidance provided by incorporated cues will direct the stem cells to the desired lineages within the multi-scale tissue structures. A single stem cell source combined with a multi-scale scaffold eliminates the need for matching organ donors, minimizes the risk of immune rejection and reduces the waiting period. The regenerative engineering toolbox that we and others are developing will be further enhanced over the next few decades leading to new strategies and mergers of disparate technologies. We will move beyond addressing individual tissue repair to complex tissue systems such as limbs, and other organ systems by capitalizing on the concept of convergence by incorporating advanced materials science, stem cell science, and developmental biology.

ACKNOWLEDGEMENTS

The authors gratefully acknowledge funding from the Raymond and Beverly Sackler Center for Biomedical, Biological, Physical and Engineering Sciences, NSF AIR 1311907 and NSF EFRI 1332329. Dr. Laurencin was the recipient of Presidential Faculty Fellow Award from the National Science Foundation.

REFERENCES

[1]   J. P. Vacanti, "Tissue engineering: the design and fabrication of living replacement devices for surgical reconstruction and transplantation," *Lancet,* vol. 354, p. S32, 1999.

[2]   Y. Fung, "A proposal to the National science Foundation for An Engineering Research Centre at USCD," *Center for the Engineering of Living Tissues. UCSD,* vol. 865023, 2001.

[3]   T. W. Bauer and G. F. Muschler, "Bone graft materials: an overview of the basic science," *Clinical orthopaedics and related research,* vol. 371, p. 10, 2000.

[4]   R. Langer and J. Vacanti, "Tissue engineering," *Science,* vol. 260, pp. 920-926, May 14, 1993 1993.

[5]   C. T. Laurencin and Y. Khan, "Regenerative Engineering," *Science Translational Medicine*, vol. 4, p. 160ed9, November 14, 2012 2012.

[6]   R. James, G. Q. Daley, and C. T. Laurencin, "Regenerative Engineering: Materials, Mimicry, and Manipulations to Promote Cell and Tissue Growth," *National Academy of Engineering - The Bridge: The Convergence of Engineering and the Life Sciences; Editors: Philip A. Sharp and Robert Langer*, vol. 43, p. 8, 2013.

[7]   W. Reichert, B. D. Ratner, J. Anderson, A. Coury, A. S. Hoffman, C. T. Laurencin, *et al.*, "2010 Panel on the biomaterials grand challenges," *Journal of Biomedical Materials Research Part A*, vol. 96, pp. 275-287, 2011.

[8]   P. A. Sharp and R. Langer, "Promoting Convergence in Biomedical Science," *Science*, vol. 333, p. 527, July 29, 2011 2011.

[9]   R. Peerani, B. M. Rao, C. Bauwens, T. Yin, G. A. Wood, A. Nagy, *et al.*, "Niche-mediated control of human embryonic stem cell self-renewal and differentiation," *EMBO J*, vol. 26, pp. 4744-55, Nov 14 2007.

[10]   T. H. Petersen, E. A. Calle, L. Zhao, E. J. Lee, L. Gui, M. B. Raredon, *et al.*, "Tissue-Engineered Lungs for in Vivo Implantation," *Science*, vol. 329, pp. 538-541, July 30, 2010 2010.

[11]   H. C. Ott, T. S. Matthiesen, S.-K. Goh, L. D. Black, S. M. Kren, T. I. Netoff, *et al.*, "Perfusion-decellularized matrix: using nature's platform to engineer a bioartificial heart," *Nature medicine*, vol. 14, pp. 213-221, 2008.

[12]   S. F. Badylak, D. Taylor, and K. Uygun, "Whole-organ tissue engineering: decellularization and recellularization of three-dimensional matrix scaffolds," *Annu Rev Biomed Eng*, vol. 13, pp. 27-53, Aug 15 2011.

[13]   T. Dvir, B. P. Timko, D. S. Kohane, and R. Langer, "Nanotechnological strategies for engineering complex tissues," *Nat Nanotechnol*, vol. 6, pp. 13-22, Jan 2011.

[14]   C. M. Kelleher and J. P. Vacanti, "Engineering extracellular matrix through nanotechnology," *J R Soc Interface*, vol. 7 Suppl 6, pp. S717-29, Dec 6 2010.

[15]   R. James, G. Kesturu, G. Balian, and A. B. Chhabra, "Tendon: Biology, biomechanics, repair, growth factors, and evolving treatment options," *Journal of Hand Surgery-American Volume*, vol. 33A, pp. 102-112, Jan 2008.

[16]   R. James, M. Deng, C. Laurencin, and S. Kumbar, "Nanocomposites and bone regeneration," *Frontiers of Materials Science*, vol. 5, pp. 342-357, 2011.

[17]   M. V. Hogan, N. Bagayoko, R. James, T. Starnes, A. Katz, and A. B. Chhabra, "Tissue engineering solutions for tendon repair," *J Am Acad Orthop Surg*, vol. 19, pp. 134-42, Mar 2011.

[18]   D. Meng, R. James, C. T. Laurencin, and S. G. Kumbar, "Nanostructured Polymeric Scaffolds for Orthopaedic Regenerative Engineering," *NanoBioscience, IEEE Transactions on*, vol. 11, pp. 3-14, 2012.

[19]   M. Borden, M. Attawia, Y. Khan, and C. T. Laurencin, "Tissue engineered microsphere-based matrices for bone repair::: design and evaluation," *Biomaterials*, vol. 23, pp. 551-559, 2002.

[20]   M. Borden, S. El-Amin, M. Attawia, and C. Laurencin, "Structural and human cellular assessment of a novel microsphere-based tissue engineered scaffold for bone repair," *Biomaterials*, vol. 24, pp. 597-609, 2003.

[21]   C. Laurencin, M. Attawia, L. Lu, M. Borden, H. Lu, W. Gorum, et al., "Poly (lactide-co-glycolide)/hydroxyapatite delivery of BMP-2-producing cells: a regional gene therapy approach to bone regeneration," Biomaterials, vol. 22, pp. 1271-1277, 2001.

[22]   N. Roveri, G. Falini, M. Sidoti, A. Tampieri, E. Landi, M. Sandri, et al., "Biologically inspired growth of hydroxyapatite nanocrystals inside self-assembled collagen fibers," Materials Science and Engineering: C, vol. 23, pp. 441-446, 2003.

[23]   T. J. Webster, C. Ergun, R. H. Doremus, R. W. Siegel, and R. Bizios, "Enhanced functions of osteoblasts on nanophase ceramics," Biomaterials, vol. 21, pp. 1803-1810, 2000.

[24]   T. J. Webster, C. Ergun, R. H. Doremus, R. W. Siegel, and R. Bizios, "Enhanced osteoclast-like cell functions on nanophase ceramics," Biomaterials, vol. 22, pp. 1327-1333, 2001.

[25]   T. J. Webster, R. W. Siegel, and R. Bizios, "Osteoblast adhesion on nanophase ceramics," Biomaterials, vol. 20, pp. 1221-1227, 1999.

[26]   Q. Lv, L. Nair, and C. T. Laurencin, "Fabrication, characterization, and in vitro evaluation of poly(lactic acid glycolic acid)/nano-hydroxyapatite composite microsphere-based scaffolds for bone tissue engineering in rotating bioreactors," Journal of Biomedical Materials Research Part A, vol. 91A, pp. 679-691, 2009.

[27]   S. G. Kumbar, R. James, S. P. Nukavarapu, and C. T. Laurencin, "Electrospun nanofiber scaffolds: engineering soft tissues," Biomedical materials, vol. 3, p. 034002, Sep 2008.

[28]   S. G. Kumbar, M. D. Kofron, L. S. Nair, and C. T. Laurencin, "Cell Behavior Toward Nanostructured Surfaces," in Biomedical Nanostructures, H. C. Gonsalves KE, Laurencin CT, Nair LS, Ed., ed Hoboken, NJ, USA: John Wiley & Sons, 2007, pp. 261-295.

[29]   W. J. Li, C. T. Laurencin, E. J. Caterson, R. S. Tuan, and F. K. Ko, "Electrospun nanofibrous structure: a novel scaffold for tissue engineering," J Biomed Mater Res, vol. 60, pp. 613-21, Jun 15 2002.

[30]   L. S. Nair, S. Bhattacharyya, J. D. Bender, Y. E. Greish, P. W. Brown, H. R. Allcock, et al., "Fabrication and optimization of methylphenoxy substituted polyphosphazene nanofibers for biomedical applications," Biomacromolecules, vol. 5, pp. 2212-20, 2004.

[31]   S. Bhattacharyya, L. S. Nair, A. Singh, N. R. Krogman, Y. E. Greish, P. W. Brown, et al., "Electrospinning of poly[bis(ethyl alanato) phosphazene] nanofibers " J Biomed Nanotechnol, vol. 2, pp. 36-45, 2006.

[32]   S. Bhattacharyya, S. G. Kumbar, Y. M. Khan, L. S. Nair, A. Singh, N. R. Krogman, et al., "Biodegradable polyphosphazene-nanohydroxyapatite composite nanofibers: scaffolds for bone tissue engineering," J Biomed Nanotechnol, vol. 5, pp. 69-75, 2009.

[33]   W. J. Li, C. T. Laurencin, E. J. Caterson, R. S. Tuan, and F. K. Ko, "Electrospun nanofibrous structure: A novel scaffold for tissue engineering," Journal of Biomedical Materials Research, vol. 60, pp. 613-621, Jun 2002.

[34]   E. M. Christenson, K. S. Anseth, J. J. van den Beucken, C. K. Chan, B. Ercan, J. A. Jansen, et al., "Nanobiomaterial applications in orthopedics," J Orthop Res, vol. 25, pp. 11-22, 2007.

[35]   L. S. Nair and C. T. Laurencin, "Nanofibers and nanoparticles for orthopaedic surgery applications," J Bone Joint Surg Am, vol. 90, pp. 128-131, 2008.

[36]   L. S. Nair, S. Bhattacharyya, and C. T. Laurencin, "Development of novel tissue engineering scaffolds via electrospinning," Expert Opin Biol Ther, vol. 4, pp. 659-668, 2004.

[37]   K. M. Woo, V. J. Chen, and P. X. Ma, "Nano-fibrous scaffolding architecture selectively enhances protein adsorption contributing to cell attachment," *J Biomed Mater Res A,* vol. 67A, pp. 531-537, 2003.

[38]   G. Pelled, K. Tai, D. Sheyn, Y. Zilberman, S. Kumbar, L. S. Nair, *et al.,* "Structural and nanoindentation studies of stem cell-based tissue-engineered bone," *J Biomech,* vol. 40, pp. 399-411, 2007.

[39]   M. T. Conconi, S. Lora, A. M. Menti, P. Carampin, and P. P. Parnigotto, "In vitro evaluation of poly[bis(ethyl alanato)phosphazene] as a scaffold for bone tissue engineering," *Tissue Eng,* vol. 12, pp. 811-9, Apr 2006.

[40]   M. Deng, S. G. Kumbar, L. S. Nair, A. L. Weikel, H. R. Allcock, and C. T. Laurencin, "Biomimetic structures: biological implications of dipeptide-substituted polyphosphazene-polyester blend nanofiber matrices for load-bearing bone regeneration " *Adv Funct Mater,* vol. 21, pp. 2641-2651, 2011.

[41]   J. A. Cooper, Jr., J. S. Sahota, W. J. Gorum, II, J. Carter, S. B. Doty, and C. T. Laurencin, "Biomimetic tissue-engineered anterior cruciate ligament replacement," *Proc Natl Acad Sci U S A,* vol. 104, pp. 3049-3054, 2007.

[42]   M. S. Peach, J. Roshan, S. T. Udaya, D. Meng, L. M. Nicole, R. A. Harry, *et al.,* "Polyphosphazene functionalized polyester fiber matrices for tendon tissue engineering: in vitro evaluation with human mesenchymal stem cells," *Biomedical materials,* vol. 7, p. 045016, 2012.

[43]   M. S. Peach, S. G. Kumbar, R. James, U. S. Toti, D. Balasubramaniam, M. Deng, *et al.,* "Design and optimization of polyphosphazene functionalized fiber matrices for soft tissue regeneration," *J Biomed Nanotechnol,* vol. 8, pp. 107-24, 2012.

[44]   R. James, U. S. Toti, C. T. Laurencin, and S. G. Kumbar, "Electrospun Nanofibrous Scaffolds for Engineering Soft Connective Tissues," *Methods Mol Biol,* vol. 726, pp. 243-258, 2011.

Mater. Res. Soc. Symp. Proc. Vol. 1621 © 2014 Materials Research Society
DOI: 10.1557/opl.2014.285

Novel and Unique Matrix Design for Osteochondral Tissue Engineering

Deborah L. Dorcemus[1,2], Syam P. Nukavarapu[1,2,3,4]
[1]Institute for Regenerative Engineering, University of Connecticut Health Center Farmington, CT 06030, U.S.A
[2]Biomedical Eng., [3]Materials Science & Eng. University of Connecticut Storrs, CT 06269, U.S.A
[4]Department of Orthopedic Surgery University of Connecticut Health Center Farmington, CT 06030, U.S.A

ABSTRACT

Osteochondral (OC) tissue is comprised of articular cartilage, the subchondral bone and the central cartilage-bone interface. To facilitate proper regeneration, an equally complex and multiphasic matrix must be used. Although mono-phasic and bi-phasic matrices were previously applied, they failed to establish the OC interface upon regeneration. In this study, we designed and developed a novel matrix with increasing pore volume from one end to other, along the matrix length. For this matrix polylactide-*co*-glycolide (PLGA) 85:15 microspheres were combined with a water-soluble porogen in a layer-by-layer fashion and thermally sintered. The resulting matrix was then porogen-leached to form a gradiently-porous structured matrix. The formation of this gradient pore structure was established using Micro-Computed Tomography (μCT) scanning. A biodegradable hydrogel was infiltrated into the structure to form a unique OC matrix where the microspheres and hydrogel phases co-exist with opposing gradients. When the individual phases are associated with osteogenic and chondrogenic growth factors, the structure-induced factor delivery might provide the spatially controlled factor delivery necessary to regenerate osteochondral tissue structure. Overall, we designed a gradient matrix system that is expected to support osteochondral tissue engineering while forming a seamless interface between the cartilage and the bone matrix.

INTRODUCTION

Osteochondral defect repair has been a significant challenge in orthopedic surgery [1,2]. Fresh or frozen allografts are commonly used as osteochondral plugs, however shortcomings associated with the allograft transplantation, such as disease transmission and lack of integration with the host tissue warrant the development of modern methods for OC defect repair and regeneration [1,3]. Tissue engineering offers several possibilities and a number of TE strategies have been recently developed for this purpose [1,4].

In the past both monophasic and biphasic matrices have been studied for OC tissue engineering [5,6,7,8]. Although, these structures somewhat supported cartilage and bone regeneration in their specified layers, they failed to establish an interface similar to that of native OC tissue [9,10]. Since OC tissue is uniquely structured, with a subchondral layer followed by three well-organized articular cartilage zones from the bottom to top of the structure, OC tissue engineering requires the development of unique matrices, which not only support regeneration of the zonal structure but also helps to establish a proper cartilage-bone interface [1,11,12].

The authors have recently reported the development of a unique matrix with continuous structure from one end to the other with the possibility of tuning the structure as desired [1]. In this paper, we demonstrated the development of a gradient matrix combined with a secondary phase that could support cartilage matrix regeneration. This matrix is unique due to the fact that the bone and cartilage supporting phases are arranged continuously along the matrix length with opposing gradients. This paper presents the fabrication and characterization aspects of this uniquely structured matrix developed for OC tissue engineering.

EXPERIMENTS

Poly (lactide-*co*-glycolide) (PLGA) 85:15 polymer was purchased from Evonik Industries (Birmingham, AL) and fabricated into microspheres using an oil-in-water emulsion process. The resulting microspheres (size range 355-425μm) were combined with different weight percents of NaCl porogen (size range 106-212μm). PLGA microspheres and porogen were brought together into a single matrix, with varying porogen content along the matrix length, using the thermal sintering and porogen leaching matrix fabrication method recently developed by the authors, Figure 1 [13]. The gradient matrix formation was determined using pore volume analysis performed via Micro-Computed Tomography (μCT).

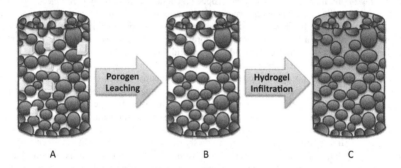

Figure 1: Schematic showing unique OC matrix fabrication: (A) Sintered PLGA matrix with increased porogen content along the matrix length, (B) Gradiently porus matrix, and (C) OC matrix with a continuous interface between the polymer and hydrogel phases.

Due to the fact that the matrix's pore volume primarily relies on the porogen content in the matrix we've determined that the gradient pore volume's increase can be introduced in a minimum of two ways: as a continuous increase or as a step-wise increase. While a step-wise gradient scaffold would contain about 5 different polymer/porogen combinations added to the sintering mold, formation of a continuously graded structure will contain additional polymer/porogen combinations subsequently layered in the chosen mold. When comparing these two designs the minimum and maximum pore volumes are expected to be similar, however, the continuously graded matrix would have a more gradual trend (Figure 2a) while the step-wise gradient matrix is expected to show steps in pore-volume, as shown in Figure 2b.

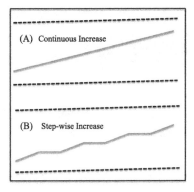

Figure 2: Representation of pore volume increase in a (a) continuously gradient matrix as well as a (b) step-wise gradient matrix

In order to determine our gradient matrix's articular cartilage and subchondral bone layers ability to support chondrogenesis as well as osteogenesis, respectively, we first fabricated each of these layers separately. Using PLGA 85:15, disk matrices (10mm diameter x 2 mm height) were prepared mimicking the first and last layers of our gradient matrix. To accomplish this, the subchondral bone matrix contained polymer, with no porogen added, and human bone marrow derived stem cells (hBMSCs) seeded directly onto it, while the articular cartilage matrix contained polymer microspheres that were 30% porogen-leached with hBMSCs embedded PuraMatrix hydrogel infiltrated into the pore spaces. The proposed PLGA-PuraMatrix is an extension of our previous work, the design of advanced "polymer-hydrogel" matrices for tissue engineering [14]. Although, PLGA and PuraMatrix gel phases were primarily chosen to demonstrate bone and cartilage regeneration, respectively, various other polymer and hydrogel phases can be selected to form gradient hybrid matrices to develop the next generation matrix systems for osteochondral tissue engineering. After 21 days in culture with osteogenic and chondrogenic media, respectively, immunofluorescence was performed to determine the presence of specific bone and cartilage markers.

DISCUSSION

Matrix pore volume Analysis

Following matrix fabrication using the step-gradient matrix method, visual and quantitative analysis with Micro CT confirms the formation of a gradiently porous matrix with an increase in pore volume from about 35 to 55 percent along the scaffold length. As seen in Figure 3, the matrix shows an increase in pore volume moving from left to right. While the microspheres within the matrix maintain a specific size range, at the lower end of the scale, the pore spaces make up the midrange of the scale all the way up to the high of 600 μm.

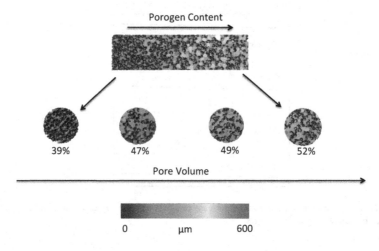

Figure 3: Matrix Pore Interconnectivity and Pore Volume via Micro-CT imaging and analysis of gradient matrices. Accessible volume space images generated by imposing specific pore diameter parameters (scale 100-600 μm) on PLGA gradient matrices (15 mm height, 5 mm diameter from a longitudinal and cross-sectional view, every 2 mm down the length of the matrix with corresponding pore volume quantification.

This gradiently porous matrix provides the framework for the osteochondral graft. Seeing as native osteochondral tissue is highly organized, top to bottom, we believe a biphasic matrix with highly specified organization of its two phases will better mimic the native tissue in that, the osteochondral matrix would also have many of the same elements throughout, but the proportions of these elements would vary top to bottom [1,11].

Matrix Characterization In Vitro

After 21 days in either osteogenic or chondrogenic media the bone and cartilage matrices were removed from culture. Immunofluorescent staining of the osteogenic as well as the chondrogenic matrices for the early markers RunX2 and Sox9, respectively, yielded positive results. These positive results in combination with the apparent cell morphology provided by the tubulin staining in Figures 4C and 4F, support the hypothesis of osteogenic and chondrogenic cell differentiation respectively. As shown in Figure 4B the cells seeded directly onto the PLGA matrix (Figure 4A), absent of hydrogel, readily expressed RunX2. Similarly, it was found that the cells embedded in the hydrogel, shown in blue (Figure 4D), were readily capable of chondrogenesis as denoted by their expression of Sox9, seen in Figure 4E. A comprehensive immunofluorescence study is underway to establish osteogenic and chondrogenic differentiation profiles along the scaffold length axis in order to demonstrate the potential of the gradient scaffold system for osteochondral tissue regeneration *in vitro*.

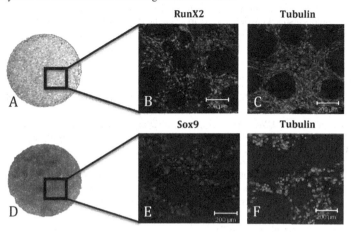

Figure 4: Human bone marrow stromal cell (hBMSC) seeded PLGA disk matrices with (A) no porogen added, cultured for 21 days in (B) osteogenic media showing positive staining for RunX2, in green, and osteogenic morphology shown using Tubulin stain in yellow (C). Disk matrices (D) 30% porogen-leached with hydrogel infiltration, seen in blue, cultured for 21 days in (E) chondrogenic media showing positive staining for Sox9, in red, and chondrogenic morphology shown using Tubulin stain in yellow (F).

RESULTS

Using a novel thermal sintering and porogen leaching method, we were able to fabricate a unique matrix with structural gradients. Micro-CT pore volume analysis showed the increase in porogen correlated with an increase in pore volume, from approximately 35 to 55 percent, therefore supporting the formation of the gradient matrix. The gradient matrix was successfully infiltrated with a biodegradable hydrogel to form a bi-phasic gradient matrix with opposing gradients of polymer to hydrogel, i.e., the bone forming and cartilage forming aspects of the matrix. The *in vitro* studies with the PLGA matrix and PLGA-hydrogel matrix seeded with hBMSCs confirms the ability of the matrices to support both osteogenesis as well as chondrogenesis, individually. These preliminary studies established the feasibility of forming a continuous yet biphasic matrix for the purpose of osteochondral tissue regeneration.

ACKNOWLEDGMENTS

Dr. Nukavarapu acknowledges support from AO Foundation (S-13-122N), NIH (AR062771), NSF (1311907) and DOD (W81XWH-11-1-0262). Ms. Dorcemus would like to thank support from NSF provided through the Northeast LSAMP Bridge to the Doctorate (BD) program (1249283). The authors also acknowledge support from the Raymond and Beverly Sackler Center for Biomedical, Biological, Physical and Engineering Sciences, The Institute for Regenerative Engineering, as well as The University of Connecticut.

REFERENCES

1. Nukavarapu SP, Dorcemus DL (2013) Osteochondral tissue engineering: Current strategies and challenges. Biotechnology Advances 31: 706-721.
2. Shao X, Goh J, Hutmacher DW, Lee EH, Zigang GE (2006) Repair of Large Articular Osteochondral Defects Using Hybrid Scaffolds and Bone Marrow-Derived Mesenchymal Stem Cells in a Rabbit Model. Tissue Eng 12.
3. Liu M, Yu X, Huang F, Cen S, Zhong G, et al. (2013) Tissue Engineering Stratified Scaffolds for Articular Cartilage and Subchondral Bone Defects Repair. Orthopedics 36: 868-873.
4. Mohan N, Dormer NH, Caldwell KL, Key VH, Berkland CJ, Detamore MS (2011) Continuous gradients of material composition and growth factors for effective regeneration of the osteochondral interface. Tissue Eng Part A 17.
5. Chu CR, Coutts RD, Yoshioka M, Harwood FL, Monosov AZ, et al. (1995) Articular-Cartilage Repair Using Allogeneic Perichondrocyte-Seeded Biodegradable Porous Polylactic Acid (PLA): A Tissue Engineering Study. Journal of Biomedical Materials Research 29.
6. Gao JZ, Dennis JE, Solchaga LA, Awadallah AS, Goldberg VM, et al. (2001) Tissue-engineered fabrication of an osteochondral composite graft using rat bone marrow-derived mesenchymal stem cells. Tissue Engineering 7.
7. Malda J, Woodfield TBF, van der Vloodt F, Wilson C, Martens DE, et al. (2005) The effect of PEGT/PBT scaffold architecture on the composition of tissue engineered cartilage. Biomaterials 26.
8. Getgood AMJ, Kew SJ, Brooks R, Aberman H, Simon T, et al. (2012) Evaluation of early-stage osteochondral defect repair using a biphasic scaffold based on a collagen-glycosaminoglycan biopolymer in a caprine model. Knee 19.
9. Schek RM, Taboas JM, Segvich SJ, Hollister SJ, Krebsbach PH (2004) Engineered osteochondral grafts using biphasic composite solid free-form fabricated scaffolds. Tissue Engineering 10.
10. Keeney M, Pandit A (2009) The Osteochondral Junction and Its Repair via Bi-Phasic Tissue Engineering Scaffolds. Tissue Engineering Part B-Reviews 15.
11. Athanasiou KA, Darling EM, Hu JC (2009) Articular Cartilage Tissue Engineering; Athanasiou K, editor: Morgan & Claypool. 168 p.
12. Marieb EN (2001) Human Anatomy & Physiology; Ueno K, editor. San Francisco: Benjamin Cummings.
13. Amini AR, Adams DJ, Laurencin CT, Nukavarapu SP (2012) Optimally porous and biomechanically compatible scaffolds for large-area bone regeneration. Tissue Eng Part A 18.

14. Igwe J, Mikael PE, Nukavarapu SP (2014) Design, fabrication and in vitro evaluation of novel polymer-hydrogel hybrid scaffold for bone tissue engineering. J Regen Med Tissue Engin. 2.

Mater. Res. Soc. Symp. Proc. Vol. 1621 © 2014 Materials Research Society
DOI: 10.1557/opl.2014.5

Reducing Infections Using Nanotechnology

Thomas J. Webster[1,2*]

[1]Department of Chemical Engineering, Northeastern University, Boston, MA, 02115, USA
[2]Center of Excellence for Advanced Materials Research, King Abdulaziz University, Jeddah, Saudi Arabia

ABSTRACT

Ventilator associated pneumonia (VAP) is a serious and costly clinical problem. Specifically, receiving mechanical ventilation for over 24 hours increases the risk of VAP and is associated with high morbidity, mortality and medical costs. Cost effective endotracheal tubes (ETTs) that are resistant to bacterial infection could help prevent this problem. The objective of this study was to determine differences in the growth of *Staphylococcus aureus* (*S. aureus*) on nanomodified and unmodified polyvinyl chloride (PVC) ETTs under dynamic airway conditions. PVC ETTs were modified to have nanometer surface features by soaking them in *Rhizopus arrhisus*, a fungal lipase. Twenty-four hour experiments (supported by computational models) showed that air flow conditions within the ETT influenced both the location and concentration of bacterial growth on the ETTs especially within areas of tube curvature. More importantly, experiments revealed a 1.5 log reduction in the total number of *S. aureus* on the novel nanomodified ETTs compared to the conventional ETTs after 24 hours of air flow. This dynamic study showed that lipase etching can create nano-rough surface features on PVC ETTs that suppress *S. aureus* growth and, thus, may provide clinicians with an effective and inexpensive tool to combat VAP.

INTRODUCTION

Despite extensive research into prevention and management, ventilator associated pneumonia (VAP) continues to be a severe, costly complication of mechanical ventilation among critically ill patients. Device related infections, such as VAP affect 28% of the patients who use mechanical ventilation. In the United States, the estimated yearly cost of hospital acquired infections (HAIs), such as VAP, can be up to 6.7 billion dollars [1]. The two pathogens most commonly associated with VAP are *Pseudomonas aeruginosa* (*P. aeruginosa*) and *Staphylococcus aureus* (*S. aureus*) [2]. VAP presents a particular intractable problem to the pediatric ICU because it is often difficult to distinguish pneumonia from other respiratory conditions common in mechanically ventilated patients [3]. This is especially true in ventilated neonates and can cause delays in targeted treatment or the prolonged use of a broad-spectrum of antibiotics leading to serious risks such as the development of multidrug resistant organisms [1].

One of the main sources of bacterial colonization within the airway is the endotracheal tube (ETT). ETTs are of special concern because these tubes provide a direct conduit from the outside environment to the lungs [4] [5]. Aggregations of microorganisms surrounded by extracellular matrix (or biofilms) on the surface of the ETT make treatment difficult as they are especially resistant to both antibiotics and the immune system of the patient.

The unique properties of nanomodified surfaces (or surfaces with features <100 nm in one direction) could provide a solution to the persistent problems of VAP [6]. Nanomaterials have unique surface energetics due to their significantly greater surface area compared to conventional, nano-smooth, materials. Such changes in surface energy undoubtedly influence initial protein interactions important for mediating bacterial adhesion [7]. The type, concentration, conformation, and bioactivity of proteins adsorbed onto a material depends on its surface chemistry, hydrophilicity, charge, topography, roughness, and energy: all of which can be influenced by nanotechnology [7]. Surface properties including roughness, stiffness and topography also effect bacterial adhesion and may play an important role in the initial stages of biofilm formation [8], [9]. Reducing bacterial adhesion on the surface of an ETT could reduce both biofilm formation and bacterial contamination within the tube.

Previous studies on VAP have concentrated on biofilm formation in straight, adult, ETTs under static conditions. However, dynamic measurements are an important part of the understanding of both biofilm formation and maintenance in the ETT and the pathogenesis of VAP. Biofilms can alter their structural properties in response to fluid effects (such as shear stress) both resisting detachment and facilitating further colonization [10]-[13]. Therefore, dynamic tests of nanomodified surfaces are necessary to characterize the full antimicrobial potential of these novel nanomodified materials, especially for ETT applications.

Due to the above, the objective of this *in vitro* study was to fabricate and test nanomodified ETTs (without the use of anti-bacterials), under simulated dynamic airflow conditions of the respiratory system, in order to determine their antimicrobial effectiveness. To achieve this, both experimental and computational analyses were performed.

EXPERIMENT

Nanomodification of ETTs

The first type of antibacterial ETT analyzed within the system was a polyvinyl chloride (PVC) ETT (Sheridan®) with nano-roughened surfaces. Nanofeatures were created on these tubes using lipases from the fungi *Rhizopus arrhisus* (Sigma Aldrich) which enzymatically degrade the PVC material. The procedure included exposing the PVC tubing to a 0.1% mass solution of *R. arrhisus* lipase dissolved in a potassium phosphate buffer at 37° C. The samples were gently agitated for a total of 48 hours and the lipase media was changed every 24 hrs.

The activity of the *R. arrhisus* lipase used in the experiment was 10.5 U/g, where one unit was defined as the amount of enzyme that catalyzed the release of 1 μmol of oleic acid per minute at pH 7.4 and 40°C. ETTs placed within the system were modified on both the inner and outer surfaces. ETTs were sterilized using ethylene oxide in a 16 hour sterilization cycle.

Material characterization

Each of the samples was coated with a gold palladium mixture to increase conductivity for scanning electron microscopy (LEO 1530VP FE-4800 Field-Emission SEM, Carl Zeiss SMT, Peabody, MA) in order to visually assess nanometer surface features. Nanoroughness was also assessed using an atomic force microscope (AFM: XE-100, Park Systems, Santa Clara, CA) under tapping mode using a 10 nm AFM tip (PPP-NCHR, Park Systems, Santa Clara, CA) with a scan rate of 0.5 Hz. Average root mean square roughness data was collected for all surfaces. Contact angle data was assessed to determine surface hydrophobicity (EasyDrop DSA20S, Kruss, Hamburg, Germany).

A BCA Protein Assay Kit (Thermo Scientific, Rockford, IL) was used to quantify the total amount of protein adsorbed on the samples. This was accomplished by soaking sterilized ETT tube pieces in TSB media for 10 hours. To test protein adsorption on the surface of the tube, ETT pieces were then removed from the TSB media and re-suspended in phosphate buffered saline with RIPA buffer, to remove adsorbed proteins. The total amount of proteins in the supernatant was then tested using the BCA assay.

Bacteria dynamic testing in a bench top airway model

Nanomodified ETTs were tested in a custom made bench top model airway adapted from the general design of Hartmann et al. (1999) [14]. ETTs with a 3.5 mm internal diameter (ID) were placed in the system. The boxes were then filled with sterile trypticase soy broth media (TSB) and placed in a water bath that was maintained at 37°C. An ETT was connected to an Infant Star 950 ventilator (Puritan Bennett, Covidien, Mansfield, MA) with positive end-expiratory pressure (PEEP) of 1 cmH$_2$O and fraction of inspired oxygen (FiO$_2$) of 0.5. Four hundred and eighty mL of TSB was inoculated with 10^3 colony forming units/milliliter (CFU/mL) of $S.$ $aureus$ (ATCC #25923). This bacterial media was then circulated into the oropharynx box over the duration of a 24 hour test, using a peristaltic pump.

At the end of each trial, ETTs were cut into ten 1.5 cm pieces. These tubes were processed using a vortex protocol to determine the number of colony forming units [15].

Air flow analysis

To fully understand the dynamic air flow conditions within this system a three dimensional computational model was created. A finite element model of the exact curvature of the ETTs was created using COMSOL Multiphysics (v4.0a, COMSOL Group, Burlington MA). Air flow through the ETT was considered to be laminar, Newtonian and incompressible. The unsteady, three dimensional form of the Navier Stokes equations were solved using the finite element method. For this flow model, the boundary conditions prescribed over the domain were as follows: i) On the solid wall of the tube a no slip condition was applied; ii) An absolute pressure of 1 atm was assumed at the outlet, as is common within the clinical setting; and iii) finally an inlet pressure (P) for a pressure limited, constant flow ventilator, was represented using a square wave, with an upper limit of (PIP-PEEP) while a lower limit of 1 atm was used [16]. Each peak was exactly 0.416 seconds wide, the inspiratory time of the ventilator.

The initial conditions for the model were as follows, i) the initial inlet velocity was set at 0 m/s and ii) the initial inlet pressure was set at 1 atm.

To gain a better understanding of the bacterial dispersion within the system, a Lagrangian particle tracing model was performed using COMSOL Multiphysics, where all particles were assumed to be a single bacterium with a diameter 1 μm [17] and a mass of 1 pg [18].

RESULTS

Material characterization

PVC ETTs cut and submersed in the lipase solution $R.$ $arrhisus$ showed visible nanofeatures. In contrast as seen below in Figure 1, control untreated PVC ETTs did not show any nanofeatures.

27

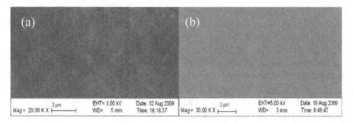

Figure 1: SEM images of nanomodified and untreated PVC ETTs: (a) Nano-R: Mag. x 20K (b) Untreated: Mag. x 30K. Nano-R was modified with *R. arrhisus*. Scale bars = 2 μm.

AFM analysis of the nanomodified ETTs confirmed SEM analysis and can be seen below. The average root mean square (RMS) roughness of the ETTs nanomodified with *R. arrhisus* was 12.5 nanometers compared to a value of 2.2 nanometers for the untreated PVC tubes; AFM scans were 5x5 μm for this analysis.

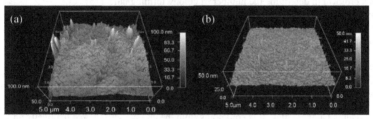

Figure 2: AFM images of PVC ETTs treated with (a) 0.1% *R. arrhisus* solution (Nano-R) for 48 hours compared to (b) untreated PVC samples.

To better understand the nanomodified surfaces, surface energy analysis was performed for the nanomodified and unmodified PVC and can be seen in Figure 3. Results showed a water contact angle of 84.4° for unmodified PVC and an angle of 68.9° for *R. arrhisus* treated (Nano-R) samples. Thus, Nano-R treatment increased hydrophilicity of the PVC.

	Untreated	Nano-R
Sample pictures		
Water contact angle	84.4	68.9

Figure 3: Increased hydrophilicity of nanomodified PVC ETTs (Nano-R) compared to controls.

The BCA assay showed an increased amount of protein adsorption on the nanomodified (Nano-R) samples compared to the untreated PVC (Figure 4), corresponding to the increased hydrophillicity of these surfaces.

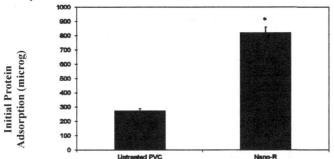

Figure 4: Increased total protein adsorption on nanomodified PVC ETTs (Nano-R) compared to controls: Nano-R was modified with *R. arrhisus* Data = mean +/- SEM; N = 3; * p < 0.01 compared to untreated.

Dynamic testing in a bench top model

A marked difference in biofilm formation on different areas of the unmodified ETTs was observed in the dynamic testing (Figure 5). Areas of ETT curvature, such as at the entrance to the mouth and the connection between the oropharynx and the larynx, were correlated with larger amounts of biofilm formation. Most importantly, colony forming units were significantly smaller (up to a 1.5 log reduction) on the nanomodified ETTs than on control ETTs, again demonstrating the efficacy of nanomodified materials as an antimicrobial without the use of pharmaceutical drugs.

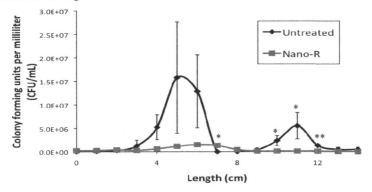

Figure 5: Dynamic lung system results for untreated versus nanomodified PVC ETTs: colony counts of *S. aureus*; N=3 (x-axis length= longitudinal ETT length), Error bars +/- 1 SE, *p < 0.05 and **p< 0.01 compared to untreated ETTs at the same time points.

Fluid analysis

Computational model results showed skewing of the air velocity profile at both curves in the ETTs, with consequent variations in the wall shear rates seen in Figure 6A. At the first curve, shear stress did vary along the lower wall, showing a weak correlation with bacterial density (correlation coefficient of about 0.37). However, the second region of curvature in the ETT showed no such relationship. This suggests that factors other than changes in shear stress (such as bacteria residence time, secondary velocity, and interactions with the wall) are important to understand differences in bacterial distribution along the ETT.

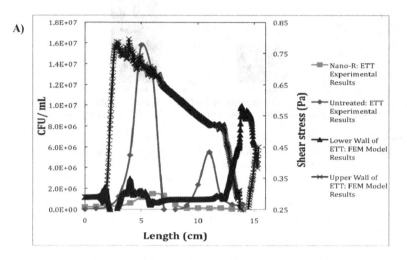

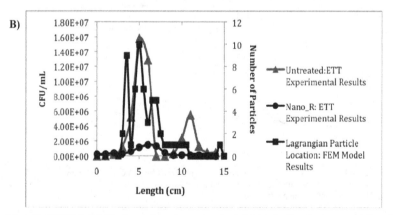

Figure 6: **A)** FEM model: Maximum shear stress versus bacterial density. Results show correlation of shear stress to increased bacteria function for the first curve. **B)** FEM model: Particle distribution versus

bacterial distribution.

Results of an associated Lagragian particle analysis can be seen in Figure 6B, where final particle (bacterium) location is graphed with bacteria distribution. A greater correlation, although still weak, between particle location and bacterial distribution can be seen within the first and second ETT curves in this model (correlation coefficient of about 0.50). Future studies will need to examine exactly why bacteria functions are decreased on certain regions of the ETT as well as on nanomodified ETTs.

CONCLUSIONS

This study showed that chemical etching techniques can create nano-rough surface features on PVC ETTs to suppress *S. aureus* growth. Compared to previous static studies [15], the present dynamic study showed a much greater decrease in bacteria function on nanomodified compared to unmodified ETTs. Dynamic air flow conditions had an effect on both the concentration and location of bacterial growth on the ETTs. *In vitro* modeling of nanomodified ETTs under dynamic conditions may further provide useful insights into their antimicrobial effectiveness. The differences in bacterial colonization recorded along the length of the tube suggest that flow and continuous contamination play a significant role in early biofilm formation. This provides evidence that static testing protocols may not reveal the important interactions at the surface of such devices *in vivo*. Most importantly, nanomodified ETT can decrease *S. aureus* growth by only changing the degree of nanofeatures of the material, not by altering chemistry or using antibiotics. These tubes could provide clinicians with a useful, inexpensive tool to combat VAP.

ACKNOWLEDGMENTS

The authors would like to thank the Hermann Foundation and NSF GK-12 for funding.

REFERENCES

1. R.W. Haley "Incidence and nature of endemic and epidemic nosocomial infections", in *Bennett and Brachman's Hospital Infections*, edited by J.V. Bennett, P.S. Brachman and W.R. Jarvis. (Lippincott Williams & Wilkins, Boston, 2007) p.359–374.
2. R. Gaynes and J.R. Edwards. Overview of nosocomial infections caused by gram-negative bacilli. Clin. Infect. Dis. **41**, 848-854 (2005).
3. Baltimore R.S. The difficulty of diagnosing VAP. Pediatrics **112**, 1420-1421(2003).
4. F.K. Bahrani-Mougeot, B.J. Paster, and S. Coleman. Molecular analysis of oral and respiratory bacterial species associated with ventilator-associated pneumonia. J. Clin. Microbiol. **45**, 1588-1593 (2007).
5. R. J. Koerner Contribution of endotracheal tubes to the pathogenesis of ventilator-associated pneumonia. J. Hosp. Infect. **35**, 83-89 (1997).
6. S. E. Carsons, *Fibronectin in Health and Disease.* (CRC Press Inc., New York,1989).
7. M.C. Machado, D. Cheng, K.M. Tarquinio, and T.J. Webster. Nanotechnology: Pediatric Applications. Pediatr. Res., **67(5)**, 500–504 (2010).

8. J. Klein Probing the interactions of proteins and nanoparticles. Proc. Natl. Acad. Sci. U.S.A. **104**, 2029-2030 (2007).

9. H. Liu, and T.J. Webster. Nanomedicine for implants: A review of studies and necessary experimental tools. Biomaterials **28**, 354-369 (2006).

10. J.A Lichter , Thompson M.T., Delgadillo M., Nishikawa T., Rubner M.F., and K. J. Van Vliet. Substrata mechanical stiffness can regulate adhesion of viable bacteria. Biomacromolecules **9(6)**,1571-1578 (2008).

11. C. Diaz, M.C Cortizo, P.L Schilardi, S.G.G Saravia, and M.A.F.L. Mele. Influence of the nano-micro structure of the surface on bacterial adhesion. Mat. Res. **10(1)**, 11-14 (2007).

12. S.A. Berger, and L. Talbot. Flow in curved pipes. Annu. Rev. Fluid Mech. **15**,461–512 (1983).

13. R. Rusconi, S. Lecuyer, L. Guglielmini, and H.A. Stone. Laminar flow around corners triggers the formation of biofilm streamers. J. R. Soc. Interface **7**,1293-1299 (2010).

14. M. Hartmann, J. Guttmann, B. Muller., T. Hallmann, and K. Geiger. Reduction of the bacterial load by the silver-coated endotracheal tube (SCET) a laboratory investigation. Technol. Health Care **7**, 359-370 (1999).

15. Seil J.T., Rubien N.M., Webster T.J., and K.M. Tarquinio. Comparison of quantification methods illustrates reduced Pseudomonas aeruginosa activity on nanorough polyvinyl chloride. J. Biomed. Mater. Res. B, 9B(1):1-7,2011.

16. R Dellinger, P Jean and S, Cinel. Regional distribution of acoustic-based lung vibration as a function of mechanical ventilation mode. Crit Care. **11**, R26 (2007).

17. *Microbiology, An Introduction.* (Tortura, Funke, Case, California, 1998).

18. Dulbecco Davis, Ginsberg Eisen, *Bacterial Physiology: Microbiology*, 2nd ed., (Harper and Row, Maryland, 1973) p.96-97.

Mater. Res. Soc. Symp. Proc. Vol. 1621 © 2014 Materials Research Society
DOI: 10.1557/opl.2014.198

Identifying Iron Oxide Based Materials that Can Either Pass or Not Pass through the *in vitro* Blood-Brain Barrier

Di Shi[1], Linlin Sun[2], Gujie Mi[1], Soumya Bhattacharya[3], Suprabha Nayar[3], Thomas J Webster[4]
[1]Department of Pharmaceutical Sciences, Northeastern University,
Boston, MA 02115, USA.
[2]Department of Bioengineering, Northeastern University,
Boston, MA 02115, USA.
[3]Materials Science and Technology Division, CSIR-National Metallurgical Laboratory,
Jamshedpur, JH 831007, India.
[4]Department of Chemical Engineering, Northeastern University,
Boston, MA 02115, USA.

ABSTRACT

In this study, an *in vitro* blood-brain barrier model was developed using murine brain endothelioma cells (b.End3 cells). By comparing the permeability of FITC-Dextran at increasing exposure times in serum-free medium to such values in the literature, we confirm that the blood-brain barrier model was successfully established. After such confirmation, the permeability of five ferrofluid (FF) nanoparticle samples, GGB (ferrofluid synthesized using glycine, glutamic acid and BSA), GGC (glycine, glutamic acid and collagen), GGP (glycine, glutamic acid and PVA), BPC (BSA, PEG and collagen) and CPB (collagen, PVA and BSA), was determined using this model. In addition, all the five FF samples were characterized by zeta potential to determine their charge as well as TEM and dynamic light scattering for determining their hydrodynamic diameter. Results showed that FF coated with collagen had better permeability to the blood-brain barrier than FF coated with glycine and glutamic acid based on an increase of 4.5% in permeability. Through such experiments, magnetic nanomaterials, such as ferrofluids, that are less permeable to the blood brain barrier can be used to decrease neural tissue toxicity and magnetic nanomaterials with more permeable to the blood-brain barrier can be used for brain drug delivery.

INTRODUCTION

As we know, the blood-brain barrier act as a sanctuary that separates somatic circulating blood from the cerebrospinal fluid in the central nervous system (CNS). Consisting of tight junctions that lie between endothelial cells in capillaries in the CNS, it keeps large or hydrophilic microorganisms and molecules from passing into the cerebrospinal fluid. Unfortunately, limited by current technologies, delivering therapeutic agents or image molecules into the brain is also blocked by these highly selective tight junctions [1-3]. However, previous research has demonstrated that small lipid-soluble-molecules which have a molecular weight less than 600 Da can be transported across the blood-brain barrier, suggesting a pathway to design novel nanoparticles which can either be inhibited or promoted to cross the blood brain barrier. Nonetheless, for neural drug delivery applications, a requirement of nanoparticle-based molecules is that they can be successfully transported across the blood-brain barrier [4]. In contrast, for whole body MRI applications, there is a requirement to keep such magnetic nanoparticles from crossing the blood brain barrier in order to decrease neural tissue toxicity.

Among several intriguing nanoparticles, superparamagnetic iron oxide nanoparticles (SPIONs) have been extensively studied because of their ability to be controlled by magnetic fields [5]. For instance, magnetic fields and nanomaterials have been identified to locate and target nucleic acids, referred to as "magnetofection", within the last decade. Moreover, magnetic nanoparticles can also be used to track brain activity and detect cellular receptors that are presented in brain diseases. SPIONs have been discovered as an excellent contrast agent which can potentially increase image contrast and improve MRI sensitivity and specificity based on their capability to significantly darken on T2-weighted and T2*-weighted images [6][7].

Under normal conditions in the body and outside of the brain, the accumulation of magnetic nanoparticles can be removed by several detoxification and antioxidant mechanisms. But if the SPIONs cross the blood brain barrier, locate inside the brain, and have not been efficiently cleared or exceed the maximum amount that cells can metabolize, the accumulation of those iron oxide nanoparticles may cause neurodegeneration and be harmful to normal brain function in the long term (even after their sugar or polymer coatings degrade). Therefore, generating innovative techniques for specifically targeting agents across the blood-brain barrier with reduced toxicity remains a challenge in the field [8].

As there are concerns about both the delivery of drugs across the blood-brain barrier and the accumulation of SPIONs in the body, this study established an in vitro blood-brain barrier model to test the ability of a new type of magnetic materials (ferrofluids) to pass through the blood-brain barrier (figure 1). In this specific case, since primary cultures of brain microvessel endothelial cells can be easily contaminated by other neurovascular cells and lose their function to tighten the blood-brain barrier; we used an immortalized cell line, b.End3 cells, in our blood–brain barrier model. We tested several variations of magnetic nanoparticles in an effort to both increase blood-brain barrier passage (for neural drug delivery applications) and decrease blood-brain barrier passage (to minimize toxicity). Results provided significant promising evidence that a combination of bioactive ligands used during in situ synthesis of ferrofluids determines whether they cross the blood brain barrier.

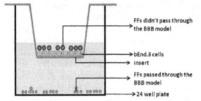

Figure 1: *In vitro* BBB model used for testing the permeability of ferrofluids.

EXPERIMENT

Material Characterization: All the five samples were characterized by zeta potential to determine their charge and dynamic light scattering for hydrodynamic diameter using Particle Size Analyzer 90 Plus; Brookhaven Instruments Company). For this, every sample was placed in a 1 ml cuvette and tested at eight separate times. Transmission electron microscopy (TEM) was also used to assess the iron oxide inner core diameter. Images were taken at 25,000x to 100,000x magnification by a JEM-1010 TEM (JEOL USA Inc.; MA) and each sample was measured over 20 times to obtain the average diameter.

Confirmation of the Blood-Brain Barrier Model: Characterization was carried out through fluorescent sugar transport to confirm the in vitro blood-brain barrier model. For the in vitro blood-brain barrier model, murine brain endothelioma (b.End3) cells (CRL–2299; American Type Culture Collection, Manassas, VA) were used, and were cultured in Dulbecco's modified Eagle medium (DMEM) with 10% fetal bovine serum (FBS) (Sigma-Aldrich) and 1% penicillin-streptomycin (Life Technologies, Carlsbad, CA). After being cultured in a sterile incubator, cells were grown to confluency in a poly-D-lysine vented cap flask (Sigma-Aldrich) before being moved to inserts. In the next step, 6.5 mm Transwell®-COL collagen-coated 0.4µm pore polytetrafluoroethylene membrane inserts were combined with 24-well plates (Sigma-Aldrich) for the model. A hemocytometer and a light microscope (Zeiss Primo Star; Peabody, MA) were used to determine the concentration of cells. After being diluted down to 105 cells per mL in DMEM + 10% FBS + 1% penicillin-streptomycin, cells in the previous cell culture section were pipetted onto the inserts. Then, the well plates with inserts were incubated daily changing medium in the outer well until cell confluency was reached. Following the desired confluency, cell culture medium was replaced with 1:1 DMEM/Ham's F12 with 1% penicillin-streptomycin for the next 96 hours.

To confirm the blood-brain barrier model, the permeability of the inserts was tested in triplicate. All the inserts were transferred to wells that contained 600 µL of Hank's Balanced Salt Solution (HBSS;

Sigma-Aldrich) before adding fluorescein isothiocyanate-labeled dextran (FITC-Dextran; molecular weight 3 kDa; Sigma-Aldrich). Immediately after the transfer, the permeability of the inserts was tested in triplicate using 100 µL of 10 µg/mL FITC-Dextran. Positive controls were set up by adding 100 µL of FITC-dextran solution with 600 µL of HBSS and negative controls were set up by adding 700 µL of HBSS only. After 2 hours of incubation, five aliquots of 100 µL were taken from each well and transferred to a blank 96-well plate with a clear bottom (Sigma-Aldrich). A fluorescent plate reader (SpectraMAX M3, Molecular Devices) was used. It was excited at 490/20 nm and measured at 528/20 nm.

Experimental Samples Tested Through the Blood-Brain Barrier Model: The previously described in vitro blood-brain barrier model was then used to test the permeability of the various nanoparticles, which were GGB (ferrofluid synthesized using 0.001% glycine, 0.001% glutamic acid and 0.01% BSA), GGC (0.001% glycine, 0.001% glutamic acid and 0.01% collagen), GGP (0.001% glycine, 0.001% glutamic acid and 0.01% PVA), BPC (0.001% BSA, 0.01% PEG and 0.01% collagen) and CPB (0.01% Collagen, 0.01% PVA and 0.001% BSA) [9]. For this, nanoparticles were diluted 1:19 with HBSS and then inserts were exposed to the nanoparticles for 2 hours. After 2 hours, a 100 µL solution was taken from each well and a full spectrum absorbance was used to determine the iron concentration that passed through the model. Each experiment was conducted in triplicate and repeated at least three times. An in vitro blood-brain barrier model based on b.End 3 cells was then used to test the permeability of the various nanoparticles. Nanoparticles were diluted 1:19 with HBSS and then inserts were exposed to the nanoparticles for 2 hours. After 2 hours, a 100 µL solution was taken from each well and a full spectrum absorbance was used to determine the iron concentration. Each experiment was conducted in triplicate and difference between means was determined using ANOVA followed by student t-tests.

DISCUSSION

Material Characterization: Dynamic light scattering revealed that the hydrodynamic diameters of the samples ranged from 20nm to 200nm (Figure 2). For zeta potential, the lowest absolute zeta value was 29mV for BPC and the highest one was 69mv for GGB (Figure 3). TEM images shows that the iron core was about 5-10nm for all of the samples (Figures 4). In addition, according to the paper published by co-author Soumya, all the five samples sharing the core-shell structure, in which iron oxide act as the core and the three coating on the outside are binded to each other by non-covalent, secondary bonds [11].

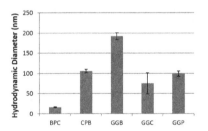

Figure 2: Average hydrodynamic diameters of samples

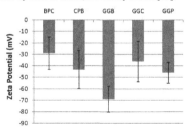

Figure 3: Average zeta potential of samples

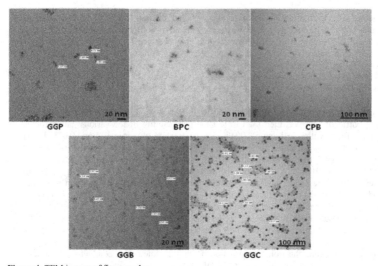

Figure 4: TEM images of five samples.
Abbreviations: GGB (ferrofluid synthesized using glycine, glutamic acid and BSA), GGC (glycine, glutamic acid and collagen), GGP (glycine, glutamic acid and PVA), BPC (BSA, PEG and collagen) and CPB (Collagen, PVA and BSA)

Confirmation of the Blood-Brain Barrier Model: The value representing permeability rate of the experimental blood-brain barrier model was in accordance with values from the literature, indicating that the model was successfully established (Figure 5) [12]. As it has been introduced by Bennett et al., the decline of permeability with increasing exposure times to serum-free medium indicated the success of the tightening of the blood-barrier model [12, 13]. To further confirm the formation of tight junction, we also took a look at the cell layers after each time point. Fluorescent images (Axio Plan 2 microscope, Zeiss) indicated that the tight junction was successfully established after 96 hours.

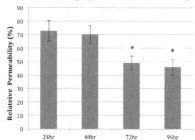

Figure 5: Permeability of fluorescein isothiocyanate-labeled dextran across the b.End3 confluent blood–brain barrier model membrane confirming establishment of an accurate blood–brain barrier. Data are shown as the mean ± SD; *P < 0.01 compared with 24hr.

Experimental Samples Tested Through the Blood-Brain Barrier Model: Results showed for the first time that the highest permeability was obtained from CPB (13.56%) and the lowest permeability was obtained from GGB (9.02%) (Figure 6). Also, ferrofluids (FF) synthesized using combination of collagen and PVA generally had higher permeability than those synthesized using glycine and glutamic acid (e.g., CPB 13.56% vs. GGB 9.02%). These results suggest that for nanoparticles that need to be delivered through the blood-brain barrier (i.e., for treating neurological diseases), FF should be coated with collagen while, on the other hand, FF should be coated with glycine and glutamic acid to keep the nanoparticles from penetrating the blood-brain barrier (i.e., for whole body MRI imaging to decrease neural tissue toxicity). In addition, in order to know the cytotoxicity that FFs have to cells, a biocompatibility experiment was set up and carried out using MTS assay. Data shows relative biocompatibility compared to control, for which cell were incubated without any sample. Results indicated that all the five samples have great biocompatibility with cells when diluted 1:19 with HBSS.

To understand whether nanoparticle diameter and/or electrokinetic potential could be used in the present study to regulate the permeability of the blood–brain barrier, a regression study was run between blood–brain barrier FF permeability and FF size, and between FF permeability and FF zeta potential. Results support the theory that large particles (e.g., GGB for192 nm) have difficulty passing through the blood-brain barrier. In addition, the correlations suggest that decreasing zeta potential (i.e., GGB for -68.95mV) could also be a good way to reduce nanoparticle permeability across the BBB. On the contrary, there is some support from the present study that increasing zeta potential (e.g., BPC for -28.99mV) or decreasing nanoparticle size (e.g., BPC for17.1nm) increased FF passage across the blood brain barrier. Further studies still need to be carried out, since the change of charge, diameter, and other factors happened simultaneously for all of the FFs in the present study, thus, making these correlations difficult to define.

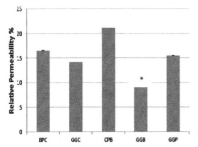

Figure 6: Blood–brain barrier permeability by sample.
Notes: Data are shown as the mean ± SD; *P < 0.01 compared with respective samples.

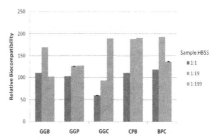

Figure 7: Cytotoxicity of five samples.
Abbreviations: GGB (ferrofluid synthesized using glycine, glutamic acid and BSA), GGC (glycine, glutamic acid and collagen), GGP (glycine, glutamic acid and PVA), BPC (BSA, PEG and collagen) and CPB (Collagen, PVA and BSA)

CONCLUSIONS

An in vitro model of blood-brain barrier was established using b.End3 cells. As the permeability decreased with increasing exposure to serum-free medium, the model was confirmed by comparing the permeability trend of FITC-dextran in serum-free medium with previous research. With the successfully established model, the permeability of five magnetic nanoparticles ferrofluids was examined. The present results suggest a possibility to manipulate magnetic nanoparticle penetration across the blood–brain barrier by control bioactive coatings. Such data lay the foundation for the modification of ferrofluids to be either coated with collagen to pass through blood-brain barrier, or to be coated with glycine and glutamic acid to avoid penetration. In addition, comparing to previous study that focused on characterizing nanoparticles with different combination of PVA, BSA, glutamic acid and collagen coating [10], we can make a further affirmation that as coating candidates, collagen or PVA itself with low concentration, or combined to glutamic acid or BSA, should have a good permeability through blood-brain barrier.

ACKNOWLEDGMENTS

The authors would like to thank Northeastern University for funding this research. One of the authors (S.B.) would like to thank Council of Scientific and Industrial Research (CSIR), Govt. of India for providing financial support in the form of Senior Research Fellowship.

REFERENCES

1. Hawkins, B.T., Davis, T.P., 2005. The blood–brain barrier/neurovascular unit in health and disease. Pharmacol. Rev. 57, 173–185.
2. Hamilton RD, Foss AJ, Leach L (2007). "Establishment of a human in vitro model of the outer blood–retinal barrier". Journal of Anatomy 211 (6): 707-16.

3. E.A. Neuwelt, et al. Engaging neuroscience to advance translational research in brain barrier biology, Nat. Rev. Neurosci. 12 (2011) 169–182.
4. Pardridge, W. M. (1995). "Transport of small molecules through the blood-brain barrier: biology and methodology." Advanced Drug Delivery Reviews 15(1-3): 5-36.
5. Seong DK, et al. Magnetic targeting of nanoparticles across the intact blood–brain barrier, Journal of Controlled Release, Volume 164, Issue 1, 28 November 2012, Pages 49-57, ISSN 0168-3659
6. Christian P, Olivier Z, Olga M. Magnetically enhanced nucleic acid delivery. Ten years of magnetofection—Progress and prospects, Advanced Drug Delivery Reviews. 2011; Volume 63, Issues 14–15:1300-1331
7. G.D. Iannetti, Richard G. Wise, BOLD functional MRI in disease and pharmacological studies: room for improvement?, Magnetic Resonance Imaging, 2007. Volume 25, Issue 6: 978-988
8. Logothetis NK, Pauls J, Augath M, Trinath T, Oeltermann A.Neurophysiological investigation of the basis of the fMRI signal. Nature. 2001;412:150–157
9. Nayar S, Sinha A, Pramanick AK. A biomimetic process for the synthesis of aqueous ferrofluids for biomedical applications. Application number 0672DEL2010. March 22, 2010
10. Dan H, Lubna S, Thomas JW. Comparison of ferrofluid and powder iron oxide nanoparticle permeability across the blood-brain barrier. Int J. of Nanomedicine. 2012; 2012:7-1.
11. S. Bhattacharya, et al. Protein-Polymer Functionalized Aqueous Ferrofluids Showing High T 2 Relaxivity. J. Biomed. Nanotechnol. 2013; Vol. 9: 1–9
12. Bennett J, et al. Blood–brain barrier disruption and enhanced vascular permeability in the multiple sclerosis model EAE. J Neuroimmunol. 2010; 229:180–191.
13. Brown RC, Morris AP, O'Neil RG. Tight junction protein expression and barrier properties of immortalized mouse brain microvessel endothelial cells. Brain Res. 2007; 1130:17–30.

Mater. Res. Soc. Symp. Proc. Vol. 1621 © 2014 Materials Research Society
DOI: 10.1557/opl.2014.344

Nanostructured Ceramic and Ceramic-Polymer Composites as Dual Functional Interface for Bioresorbable Metallic Implants

Ian Johnson[1], Qiaomu Tian[1], Huinan Liu[1,2]
[1]Department of Bioengineering, [2]Materials Science and Engineering, University of California, Riverside, CA 92521, U.S.A.

ABSTRACT

Millions of medical implants and devices (e.g., screws, plates, and pins) are used each year worldwide in surgery, and traditionally the components have been limited to permanent metals (e.g., stainless steel, titanium alloys) and polyester-based absorbable polymers. Because of clinical problems associated with these traditional materials, a novel class of biodegradable metallic materials, i.e., magnesium-based alloys, attracted great attention and clinical interests. Magnesium (Mg) is particularly attractive for load-bearing orthopedic applications because it has comparable modulus and strength to cortical bone. Controlling the interface of Mg with the biological environment, however, is the key challenge that currently limits this biodegradable metal for broad applications in medical devices and implants. This paper will particularly focus on creating nanostructured interface between the biodegradable metallic implant and surrounding tissue for the dual purposes of (1) mediating the degradation of the metallic implants and (2) simultaneously enhancing bone tissue regeneration and integration. Nanophase hydroxyapatite (nHA) is an excellent candidate as a coating material due to its osteoconductivity that has been widely reported. Applying nHA coatings or nHA containing composite coatings on Mg alloys is therefore promising in serving the needed dual functions. The composite of nHA and poly(lactic-co-glycolic acid) (PLGA) as a dual functional interface provides additional benefits for medical implant applications. Specifically, the polymer phase promotes interfacial adhesion between the nHA and Mg, and the degradation products of PLGA and Mg neutralize each other. Our results indicate that nHA and nHA/PLGA coatings slow down Mg degradation rate and enhance adhesion of bone marrow stromal cells, thus promising as the next-generation multifunctional implant materials. Further optimization of the coatings and their interfacial properties are still needed to bring them into clinical applications.

INTRODUCTION

Magnesium (Mg) possesses many desirable properties for use as a biomaterial. Mg is non-toxic [1], biodegradable [1], neuroprotective [2], antibacterial [3], and has anticancer properties [4]. Mg can enhance osteointegration of orthopedic implants because it is osteoconductive and improves bone growth around implants *in vivo* [5,6]. Mg has similar mechanical properties to cortical bone, which enables it to be used for load bearing applications while minimizing stress shielding [7]. These desirable properties make Mg a promising material for biodegradable implants that serve temporary functions in the body, such as internal fixators [8] or stents [3]. However, Mg degrades too rapidly *in vivo* to be used for most implants [9]. The galvanic degradation reactions release hydroxide ions (OH⁻) and hydrogen gas (H_2) [10] as the degradation products. The OH⁻ may increase the local pH to a cytotoxic level if the degradation is too fast [11]. The H_2 can form gas cavities in the surrounding tissues and delaminate the protective coatings [12]. Finally, rapid Mg degradation can lead to premature mechanical failure of implants. Thus, it is critical to slow down Mg degradation.

Surface coatings are often used to control the rate of Mg degradation by protecting Mg from direct exposure to the aggressive ions in the physiological fluids [13]. Controlling Mg degradation rates will reduce the release rate of degradation products, and preserve the mechanical properties of Mg substrates for the desired period. Ideally, the coatings should limit the degradation of Mg implants to a rate near that of tissue growth [14,15]. Hydroxyapatite (HA) is a phase of calcium phosphate (CaP) that is commonly used to coat orthopedic implants and devices [16]. HA is particularly beneficial to orthopedic implants because it mimics the mineral content of natural bone tissue, which provides an osteoconductive surface. Nanophase HA (nHA) has even better osteoconductivity than conventional micron-sized HA [17].

However, significant challenges remain for the clinical translation of HA coatings for Mg based bioresorbable metallic implants. The thermal expansion coefficient mismatch between HA and the metal substrate is a major source of cracks in the HA coating [18]. This challenge is especially prevalent during sintering processes due to shrinkage of the HA coating [19], and residual stresses can also damage the coating later [18]. The cracks in the coating allow the transport of aggressive ions and electrolyte to the surface of the Mg substrate to induce degradation. Many low temperature HA coating processes create porous structures in the coatings, which will likewise lead to rapid localized Mg degradation. There are other challenges with current HA coating processes. High temperature HA coating processes can alter the grain size and crystallinity of HA, or convert it into a different phase of calcium phosphate [20]. Low temperature HA coating processes often create coatings with low density or weak adhesion to their substrates. In order to control Mg degradation and tissue integration successfully, both the challenges in the materials themselves and their coating processes must be resolved. Developing a biocomposite coating composed of nHA and poly(lactic-co-glycolic acid) (PLGA) on Mg provides a promising solution in improving the coating efficacy in comparison with nHA coating alone.

EXPERIMENTS
Hydroxyapatite (HA) Coatings
IonTite, patented technology by Spire Biomedical, is a room temperature coating process that has been used to coat many materials with HA. A major advantage of IonTite is that it can preserve the chemistry and crystal size of nanophase HA (nHA) during the coating process [21]. Therefore, nHA coating was deposited on Mg by this technology.
Nanophase HA (nHA) and PLGA Composite Coatings
The thermal expansion coefficient mismatch remains a challenge for coating processes that form direct interfaces between two very different materials. This challenge may be addressed by suspending HA within a flexible polymer matrix and then depositing the suspension onto Mg substrates. The polymer matrix provides a flexible interface between Mg and HA that minimizes cracking. The polymer matrix also enables HA coatings to be deposited at room temperature. Poly(lactic co glycolic acid) (PLGA) is frequently used as the matrix for HA composites in scaffolds [22,23] and coatings [24]. PLGA is of interest because it is one of the few FDA approved polymers for biodegradable implants, and the acidic PLGA degradation products may neutralize the alkaline Mg degradation products. PLGA has low osteoconductivity by itself, but suspension of nHA within it will result in a highly osteoconductive composite material [22,23]. The osteoconductivity of the composite is enhanced when the HA is nanophase and well dispersed throughout the polymer matrix [25]. A ratio of 30% HA to 70% polymer provides the optimum

balance of mechanical and biological properties [25], and was therefore used for the composite coatings on Mg. The synergy between Mg, nHA, and PLGA makes them an attractive combination for orthopedic applications.

(1) Spin Coating of nHA/PLGA Composite

Spin coating is a simple process that offers excellent control over coating uniformity and thickness. Thus, nHA/PLGA was spin coated on Mg substrates.

(2) Electrophoretic Deposition of nHA/PLGA Microspheres

The spin coating process has several limitations. Only flat substrates, preferably circles, may be spin coated. However, most orthopedic implants have complex geometries. The spin coating process also wastes a large amount of coating material, which increases production costs and environmental impact. Electrophoretic deposition (EPD) is an alternative coating process that uses an electric field to make particles migrate to the substrate to be deposited. This alternative coating process may coat complex geometries with less wasted material than spin coating [19]. Thus, nHA/PLGA microspheres were prepared *via* the emulsion method and then electrophoretically deposited onto Mg substrates.

RESULTS AND DISCUSSION

IonTite deposited nHA coatings significantly reduced the mass loss rates of Mg substrates immersed in revised simulated body fluid (rSBF), reduced $[Mg^{2+}]$ in the media after immersion, and minimized alkalization of the media (Figure 1) [11]. The initial similarity of the media $[Mg^{2+}]$ and pH for the two different sample types was due to the initial coating fixture. Mg substrate was not completely covered by HA coating; and the four corners were not coated due to the coating fixture design (Figure 1A). This will be addressed in the future coating optimization process. The non-coated corners degraded rapidly until reaching a pH where degradation was no longer thermodynamically favored, and so the sample mass loss and media pH of the nHA coated Mg substrates were nearly identical to the non-coated Mg substrates during the first 24 hours of immersion. The nHA coating demonstrated its effectiveness for days 3-17 of the immersion study because most of the non-coated corners had already dissolved. After 19 days immersion, the $[Mg^{2+}]$ and pH for the non-coated samples were lower than the nHA coated samples because the non-coated samples had been almost completely degraded. For the nHA coated Mg substrates, it took 31 days to degrade, as opposed to 21 days for the non-coated Mg substrates. EDX analysis showed that the nHA coated Mg substrates still had a very small percentage of Mg on the surface after 17 days of immersion in rSBF, demonstrating that the nHA coating was stable. All of these results indicate that the nHA coating provided significant control over the Mg degradation rate. The nHA coating had no statistically significant effect on the adhesion of bone marrow stromal cells (BMSC). As the BMSC adhesion study only lasted for 24 hours, both coated and non-coated samples had a similar alkaline media pH. This study demonstrated that complete coating coverage of Mg substrates with nHA is necessary for improvement of their degradation rate and osteoconductivity.

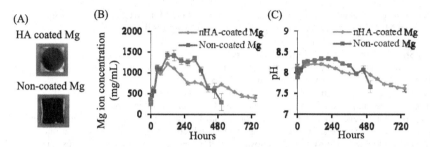

Figure 1: (A) Photographs of HA coated and non-coated Mg before immersion study. (B) Mg^{2+} ion concentration in the rSBF after immersion of nHA coated and non-coated Mg samples. (C) pH of rSBF after immersion of nHA coated and non-coated Mg samples.

A nHA/PLGA coating was deposited onto Mg substrates by spin coating, which significantly reduced the Mg degradation rate in rSBF during subsequent Tafel tests [12]. However, during immersion tests in rSBF the nHA/PLGA coatings only minimized $[Mg^{2+}]$ and alkalosis during the initial time points. The nHA/PLGA coatings became less effective over time as their structure and interfaces with the Mg substrates were undermined by H_2 evolution, eventually leading to delamination (Figure 2). Water had diffused through the PLGA matrices to reach the Mg substrates, where Mg degradation generated H_2. The delamination of the coatings demonstrated that water diffusion through coatings must be reduced and that the coating structure and adhesion strength must be improved. Interestingly, nHA/PLGA coatings had greater calcium deposition in rSBF than PLGA coatings, which implies that the nHA/PLGA composite coatings improved the osteoconductivity of the surface. The nano-scale morphology of the coating (Figure 3A) could also improve osteoconductivity, which will be confirmed in the future *in vitro* and *in vivo* studies. Biomimetically deposited CaP coatings [26] were used as a control, and provided limited protection to Mg substrates during the initial time points. The CaP coatings also had reduced effectiveness at later time points because the CaP coatings lacked a compact structure or strong adherence to their substrates.

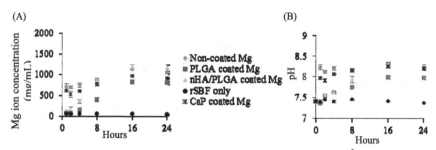

Figure 2: Immersion of coated and non-coated Mg in rSBF. (A) Dissolved $[Mg^{2+}]$ and (B) pH of media after immersion of samples.

The EPD process resulted in the composite coatings (Figure 3C) with an entirely different microstructure when compared with the coatings deposited by spin coating method (Figure 3A). The EPD deposited nHA/PLGA coatings had abundant nano-scale morphology and numerous pores. The pores can be sealed by annealing or melt heat treatments. Alternatively, the porosity of the EPD coatings may improve osteointegration if used as the outer layer of a multiple layer coating. The porosity of scaffolds is known to play important roles in osteointegration [27].

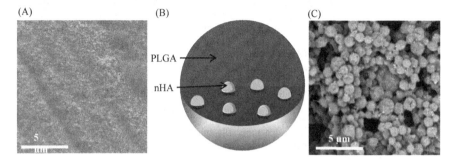

Figure 3: (A) SEM image of the nHA/PLGA composite deposited on Mg substrate using spin coating. (B) Schematic of nHA/PLGA microspheres that were prepared by the emulsion method and used in EPD process. (C) SEM image of nHA/PLGA submicron spheres deposited on Mg substrate using EPD process.

CONCLUSIONS

HA will likely be an integral part of the clinical translation of Mg based bioresorbable orthopedic implants due to its excellent osteoconductivity and versatility as a coating material. Nanophase HA and polymer composite coatings are more likely to be successful than single phase HA coatings because of the thermal expansion coefficient mismatch between HA and Mg. The nHA/polymer composites are promising because the polymer can act as a flexible matrix and binding agent to hold nHA to a substrate, which prevents many of the complications of a direct HA-Mg interface. Further optimization of the coatings and their interfacial properties are still needed to bring them into clinical applications.

ACKNOWLEDGMENTS

This material is based upon study supported by the National Science Foundation under Grant CBET BRIGE 1125801. Any opinions, findings, and conclusions or recommendations expressed in this material are those of the authors and do not necessarily reflect the views of the National Science Foundation. The authors would like to thank the Central Facility for Advanced Microscopy and Microanalysis (CFAMM) at the University of California at Riverside for access of Scanning Electron Microscope (SEM).

REFERENCES

1. Zhang Y, Ren L, Li M, Lin X, Zhao HF, Yang K. Preliminary Study on Cytotoxic Effect of Biodegradation of Magnesium on Cancer Cells. Journal of Materials Science & Technology 2012;28(9):769-772.

2. Saver JL, Kidwell C, Eckstein M, Starkman S, Investigators F-MPT. Prehospital neuroprotective therapy for acute stroke - Results of the field administration of stroke therapy-magnesium (FAST-MAG) pilot trial. Stroke 2004;35(5):E106-E108.

3. Lock JY, Wyatt E, Upadhyayula S, Whall A, Nunez V, Vullev VI, Liu H. Degradation and antibacterial properties of magnesium alloys in artificial urine for potential resorbable ureteral stent applications. Journal of Biomedical Materials Research Part A 2013.

4. Nan M, Yangmei C, Bangcheng Y. Magnesium metal - a potential biomaterial with anti-bone cancer properties. Journal of Biomedical Materials Research Part A 2013.

5. Park J-W, Kim Y-J, Jang J-H, Song H. Osteoblast response to magnesium ion-incorporated nanoporous titanium oxide surfaces. Clinical Oral Implants Research 2010;21(11):1278-1287.

6. Janning C, Willbold E, Vogt C, Nellesen J, Meyer-Lindenberg A, Windhagen H, Thorey F, Witte F. Magnesium hydroxide temporarily enhancing osteoblast activity and decreasing the osteoclast number in peri-implant bone remodelling. Acta Biomaterialia 2010;6(5):1861-8.

7. Brar HS, Platt MO, Sarntinoranont M, Martin PI, Manuel MV. Magnesium as a biodegradable and bioabsorbable material for medical implants. Jom 2009;61(9):31-34.

8. Erdmann N, Angrisani N, Reifenrath J, Lucas A, Thorey F, Bormann D, Meyer-Lindenberg A. Biomechanical testing and degradation analysis of MgCa0.8 alloy screws: A comparative in vivo study in rabbits. Acta Biomaterialia 2011;7(3):1421-1428.

9. Witte F. The history of biodegradable magnesium implants: A review. Acta Biomaterialia 2010;6(5):1680-1692.

10. Song G, Atrens A. Understanding Magnesium Corrosion—A Framework for Improved Alloy Performance. Advanced Engineering Materials 2003;5(12):837-858.

11. Iskandar ME, Aslani A, Liu HN. The effects of nanostructured hydroxyapatite coating on the biodegradation and cytocompatibility of magnesium implants. Journal of Biomedical Materials Research Part A 2013;101A(8):2340-2354.

12. Johnson I, Akari K, Liu H. Nanostructured Hydroxyapatite/Poly(lactic-co-glycolic acid) (PLGA) Composite Coating for Controlling Magnesium Degradation in Simulated Body Fluid. Nanotechnology 2013;24(37).

13. Johnson I, Liu HN. A Study on Factors Affecting the Degradation of Magnesium and a Magnesium-Yttrium Alloy for Biomedical Applications. PLoS One 2013;8(6).

14. Hing KA, Wilson LE, Buckland T. Comparative performance of three ceramic bone graft substitutes. Spine Journal 2007;7(4):475-490.

15. Sung HJ, Meredith C, Johnson C, Galis ZS. The effect of scaffold degradation rate on three-dimensional cell growth and angiogenesis. Biomaterials 2004;25(26):5735-42.

16. Guan RG, Johnson I, Cui T, Zhao T, Zhao ZY, Li X, Liu HN. Electrodeposition of hydroxyapatite coating on Mg-4.0Zn-1.0Ca-0.6Zr alloy and in vitro evaluation of degradation, hemolysis, and cytotoxicity. Journal of Biomedical Materials Research Part A 2012;100A(4):999-1015.

17. Ergun C, Liu HN, Halloran JW, Webster TJ. Increased osteoblast adhesion on nanograined hydroxyapatite and tricalcium phosphate containing calcium titanate. Journal of Biomedical Materials Research Part A 2007;80A(4):990-997.
18. Evans AG, Crumley GB, Demaray RE. On the Mechanical-Behavior of Brittle Coatings and Layers. Oxidation of Metals 1983;20(5-6):193-216.
19. Wei M, Ruys AJ, Swain MV, Kim SH, Milthorpe BK, Sorrell CC. Interfacial bond strength of electrophoretically deposited hydroxyapatite coatings on metals. Journal of Materials Science-Materials in Medicine 1999;10(7):401-409.
20. Wang BC, Chang E, Lee TM, Yang CY. Changes in Phases and Crystallinity of Plasma-Sprayed Hydroxyapatite Coatings under Heat-Treatment - a Quantitative Study. Journal of Biomedical Materials Research 1995;29(12):1483-1492.
21. Sato M, Sambito MA, Aslani A, Kalkhoran NM, Slamovich EB, Webster TJ. Increased osteoblast functions on undoped and yttrium-doped nanocrystalline hydroxyapatite coatings on titanium. Biomaterials 2006;27(11):2358-2369.
22. Lock J, Nguyen TY, Liu HN. Nanophase hydroxyapatite and poly(lactide-co-glycolide) composites promote human mesenchymal stem cell adhesion and osteogenic differentiation in vitro. Journal of Materials Science-Materials in Medicine 2012;23(10):2543-2552.
23. Lock J, Liu HN. Nanomaterials enhance osteogenic differentiation of human mesenchymal stem cells similar to a short peptide of BMP-7. International Journal of Nanomedicine 2011;6:2769-2777.
24. Ravichandran R, Ng CCH, Liao S, Pliszka D, Raghunath M, Ramakrishna S, Chan CK. Biomimetic surface modification of titanium surfaces for early cell capture by advanced electrospinning. Biomedical Materials 2012;7(1).
25. Liu H, Webster T. Enhanced biological and mechanical properties of well-dispersed nanophase ceramics in polymer composites: from 2D to 3D printed structures. Materials Science and Engineering 2011;31(2):77-89.
26. Zhang YJ, Zhang GZ, Wei M. Controlling the Biodegradation Rate of Magnesium Using Biomimetic Apatite Coating. Journal of Biomedical Materials Research Part B-Applied Biomaterials 2009;89B(2):408-414.
27. Karageorgiou V, Kaplan D. Porosity of 3D biomaterial scaffolds and osteogenesis. Biomaterials 2005;26(27):5474-91.

Mater. Res. Soc. Symp. Proc. Vol. 1621 © 2014 Materials Research Society
DOI: 10.1557/opl.2014.199

Novel Silicone-Epoxy Composites for Dental Restorations

Liyun Ren[1], Vaibhav Pandit[2], Amanda Mixon[2], Crivello James[3], Shiva P. Kotha[2]

[1]Materials Science and Engineering, Rensselaer Polytechnic Institute, 110 8th Street, Troy, NY 12180, U.S.A.
[2]Biomedical Engineering, Rensselaer Polytechnic Institute, 110 8th Street, Troy, NY 12180, U.S.A.
[3]Chemistry and Chemical Biology, Rensselaer Polytechnic Institute, 110 8th Street, Troy, NY 12180, U.S.A.

ABSTRACT

A novel silicone-epoxy oligomer was synthesized and evaluated for its use as the polymer in photopolymerizable dental composites. This synthesized oligomer contained rigid and non-rigid groups with 1-8 epoxy functionalities as characterized using ^{1}H NMR and ^{29}Si NMR and MALDI-TOF analysis. In comparison to the traditional BisGMA/TEGDMA monomer system, the photo-polymerized silicone-epoxy demonstrated significantly improved material properties (148% greater elastic modulus, 12% greater ultimate strength, 48% greater fracture toughness), as well as 61% lower polymerization shrinkage and 58% lower polymerization stresses. Furthermore, the silicone-epoxy system demonstrated enhanced resistance to degradation of its material properties after accelerated (24hr) aging, i.e. exposure to severe hydrolytic (boiling in ethanol at 100 □C), oxidative (exposed to 5% H_2O_2), and low pH (0.05M acetic acid) stress. Under these conditions, the properties of conventional BisGMA/TEGDMA systems deteriorated by 22-47%, while the properties of the silicone-epoxy composites decreased by 2-10%. Bisphosphonate additives enhanced the precipitation of mineral in a dose-dependent manner, but inhibited polymerization due to interactions with epoxy groups. Bisphosphonate additives also dose-dependently demonstrated anti-bacterial efficacy as demonstrated using live-dead, MTT and crystal violet assays. The silicone-epoxy polymer was demonstrated to be biocompatible when compared to tissue culture plastic. When calcium fluoride was incorporated into this system, fluoride was found to be released quantities significant enough to engender anti-bacterial effects. In summary, the designed multifunctional dental resin exhibits higher stability as demonstrated by lower chemical, mechanical and enzymatic degradation.

INTRODUCTION

Most composites used in the clinic to repair cavities consist of a mixture of bisphenol-acrylate resin mixed predominantly with silane coated silica based particulate, along with other additives. In the short term, one of the main drawbacks of using these bisphenol-acrylate based resin composites are the stresses associated with volumetric shrinkage [1-3]. This stress, termed the polymerization stress or shrinkage stress, can directly or indirectly lead to failure of the restoration [4,5]. Over the long-term, there are concerns related to hydrolytic, oxidative, and enzymatic degradation of the polymer and release of the unreacted monomer or its by-products [6-10]. Furthermore, since the FDA has banned the use of bisphenol containing materials in some plastics based on concerns related to their estrogen mimicry [11], there is an increasing push to remove this component from photo-polymerizable dental composites.

Of the many different polymers that are less susceptible to degradation in the oral micro-environment (where they are exposed to bacterial, enzymatic, oxidation, pH, temperature, and other), silicone-epoxy polymers are among the most robust for dental applications [12-14]. Silicone-epoxies are also more biocompatible in comparison to acrylate based materials [15-17]. As a consequence, cationic photo-polymerization of silicone-epoxies have been investigated for use in dental composites and one such silicone-epoxy monomer is currently used in the clinic (Silorane from 3M) [18,19]. However, this commercial product exhibits drawbacks related to lower rate of polymerization and higher heat generation without significant benefits in material properties or shrinkage stress in comparison to traditional bisphenol-acrylate based polymers [20]. One mechanism to overcome these drawbacks is to replace the monomer with an oligomer. So, the goal of this study is to evaluate a novel silicone-epoxy oligomer for use in dental composites. We further investigate the use of bisphosphonates to precipitate mineral and their use as anti-bacterial agents within the composite. Finally, we investigate the extent of fluoride release from calcium fluoride particulate incorporated into the polymer.

EXPERIMENT

The silicone-epoxy oligomer was synthesized by refluxing 2-(3,4-epoxycyclohexylethyl) trimethoxy-silane over an ion-exchange resin at 60 °C in the presence of a solvent [21]. The oligomer was filtered and characterized by mass spectroscopy (MALDI-TOF) as well as with ^{1}H and ^{29}Si NMR [22]. Photo-polymerization of the oligomer was carried using a cationic initiator with anthracure as a photo-sensitizer with a 500 mW/cm^2 dental light source (peak emission at 470nm). Properties of the novel oligomer were compared to those based on bisphenol-acrylate based polymers (50:50 mix of BisGMA/TEGDMA with camphorquinone and amine-based initiator). Measurements of polymerization shrinkage, polymerization stress were performed. Mechanical properties were determined using 3-point bending and mode I fracture toughness tests. Mechanical properties were also obtained after samples were exposed to diverse environments (boiling in 100% ethanol, exposure to low pH – 0.05M acetic acid, and exposure to 5% peroxide). Biocompatibility of the silicone-epoxy was assayed by comparing proliferation of osteoblasts and fibroblasts plated on discs of photo-polymerized silicone-epoxy, relative to proliferation of these cells on tissue culture plastic. Fluoride efflux from photo-polymerized composites of calcium fluoride nano-particulate containing silicone-epoxy oligomer was assayed using atomic absorption and the concentration related to anti-bacterial efficacy

Silicone-epoxy oligomer was mixed with different contents of a bisphosphonate (0, 0.1, 0.5 and 1 wt% etidronate) and this was evaluated for anti-bacterial efficacy as well as the propensity to induce mineral precipitation. Discs were plated in media supporting Streptococcus Mutans. Anti-bacterial efficacy due to the extent of etidronate incorporation was assayed by live-dead staining, MTT assays, and with crystal violet staining. Propensity for mineralization was assayed by soaking discs with different amounts of etidronate in calcium-phosphate containing simulated dental fluid for 7 days. Samples were then removed and the amount of calcium-phosphate mineral deposition was assayed using SEM and EDAXS.

RESULTS AND DISCUSSION

Mass spectroscopy results indicated the presence of oligomers with bulky epoxy containing side-groups hanging off a linear chain with one set of oligomers containing bulky end-groups as seen in figure 1. These results were supported by ^{1}H and ^{29}Si NMR.

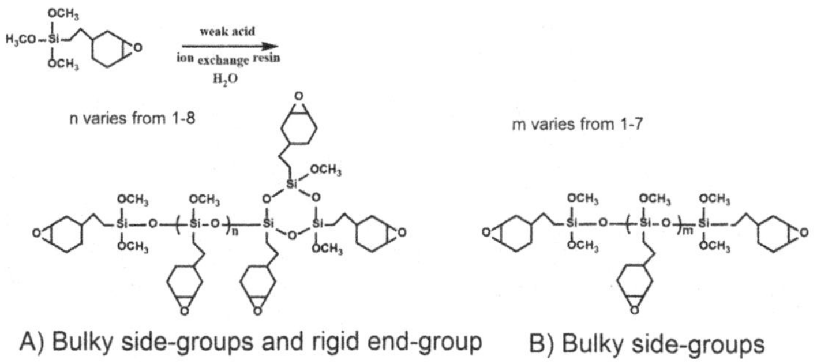

A) Bulky side-groups and rigid end-group   B) Bulky side-groups

Figure 1. Mass spectroscopy studies indicated two different products were generated. These results were supported by NMR.

Shrinkage and Shrinkage Stress

Shrinkage determined using Watts bonded disc method [23] was 2.3 ± 0.3% for silicone oligomer and 5.4 ± 0.7% for BisGMA/TEGDMA. Shrinkage stress, determined using a rigid frame holding a fixed displacement, was 3.0 ± 0.17 MPa for the silicone oligomer mix compared to 7.1 ± 0.38 MPa for BisGMA/TEGDMA . The reduction is probably due to the multiple ring-opening oxirane groups present in each oligomer.

Material Properties, Biocompatibility and Fluoride efflux

Material properties (such as elastic modulus, ultimate strength and fracture toughness) of the silicone-epoxy polymer were superior to bisphenol-acrylate polymer at 4hrs, 24hrs, and 72hrs after photo-polymerization. Elastic modulus improved rapidly from 3.6 ± 0.37 GPa (1.48 ± 0.23 GPa for bisphenol-acrylate) at 4 hrs after photo-polymerization to 5.4 ± 0.4 GPa (2.2 ± 0.23 GPa for bisphenol-acrylate) at 72 hrs. Similarly strength improved from 85.9 ± 6.6 MPa (71.9 ± 6.0 MPa for bisphenol-acrylate) at 4hrs to 107.0 ± 6. MPa (95.4 ± 8.9 MPa for bisphenol-acrylate) at 72hrs. Fracture toughness also improved from 1.36 ± 0.08 (0.82 ± 0.07 for bisphenol-acrylate) at 4hrs to 1.62 ± 0.07 (1.09 ± 0.04 for bisphenol-acrylate) at 72hrs.

Upon exposure to various insults (boiling in ethanol, exposure to low pH, and exposure to peroxide), the material properties of the silicone-epoxy samples degraded from 2%-10% while the material properties of the bisphenol-acrylate samples degraded from 22%-47%.

Osteoblasts and fibroblasts indicated increased proliferation on silicone-epoxy in comparison to tissue culture plastic. Proliferative capability for osteoblasts was greater than that of fibroblasts on the silicone-epoxy polymer discs.

The concentration of fluoride released from 5wt% calcium fluoride nano-particulate incorporated into silicone-epoxy oligomer (0.2mM), is sufficient to generate anti-bacterial efficacy.

Mineralization and anti-bacterial efficacy of etidronate

Increasing concentrations of etidronate in the silicone-epoxy oligomer retarded photo-polymerization, but caused higher calcium-phosphate mineralization on discs exposed to simulated dental fluid containing calcium-phosphates.

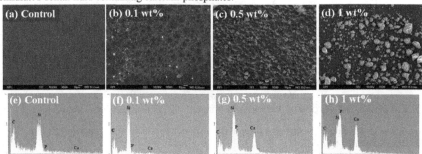

Figure 2. Samples containing increasing amounts of etidronate demonstrate greater mineralization on the samples. Top, left to right, backscatter SEM images of samples as etidronate increases from 0% (a), 0.1% (b), 0.5% (c) to 1% (d). Bottom, left to right, EDAXS images of samples indicating higher calcium and phosphorous in samples with increasing amounts of etidronate.

Results from live-dead staining, MTT assay and the crystal violet staining indicate that the highest concentrations of etidronate used (1 wt%) demonstrated the greatest anti-bacterial efficacy with 0.5 wt% etidronate demonstrating smaller effects and no discernible effects at 0.1 wt% etidronate.

CONCLUSIONS

This study demonstrates the suitability of a novel silicone-epoxy oligomer for use as the polymer in dental composites. Furthermore, additives that increase mineralization and can generate anti-bacterial efficacy were also investigated. Further studies need to be conducted to determine the extent to which material properties are enhanced with the addition of silica or other reinforcements. Fatigue and wear studies need to be conducted on the resin and on the resin-silica composite. A more comprehensive suite of biocompatibility and mutagenesis assays need to be conducted before the resin is determined to be suitable for the clinic.

ACKNOWLEDGMENTS

The authors would like to thank Ms S. Singnisai in making samples. Research was supported by NSF CMMI 1200270.

REFERENCES

(1)   Braga, R. R.; Ballester, R. Y.; Ferracane, J. L. *Dent Mater* **2005**, *21*, 962.
(2)   Davidson, C. L.; Feilzer, A. J. *J Dent* **1997**, *25*, 435.
(3)   Kleverlaan, C. J.; Feilzer, A. J. *Dent Mater* **2005**, *21*, 1150.
(4)   Choi, K. K.; Ferracane, J. L.; Ryu, G. J.; Choi, S. M.; Lee, M. J.; Park, S. J. *Oper Dent* **2004**, *29*, 462.
(5)   Perdigao, J.; Duarte, S.; Gomes, G. *J Adhes Sci Technol* **2009**, *23*, 1201.
(6)   Koin, P. J.; Kilislioglu, A.; Zhou, M.; Drummond, J. L.; Hanley, L. *J Dent Res* **2008**, *87*, 661.
(7)   Ferracane, J. L. *Dent Mater* **2006**, *22*, 211.
(8)   Ferracane, J. L. *Dent Mater* **2011**, *27*, 29.
(9)   Seiss, M.; Langer, C.; Hickel, R.; Reichl, F. X. *Arch Toxicol* **2009**, *83*, 1109.
(10)   Yourtee, D. M.; Smith, R. E.; Russo, K. A.; Burmaster, S.; Cannon, J. M.; Eick, J. D.; Kostoryz, E. L. *J Biomed Mater Res* **2001**, *57*, 522.
(11)   FDA; Administration, U. S. F. a. D., Ed. 2013; Vol. http://www.fda.gov/%20newsevents/publichealthfocus/ucm064437.htm.
(12)   Porto, I. C.; de Aguiar, F. H.; Brandt, W. C.; Liporoni, P. C. *J Dent* **2013**, *41*, 732.
(13)   Arocha, M. A.; Mayoral, J. R.; Lefever, D.; Mercade, M.; Basilio, J.; Roig, M. *Clin Oral Investig* **2013**, *17*, 1481.
(14)   Borges, A. F.; Santos Jde, S.; Ramos, C. M.; Ishikiriama, S. K.; Shinohara, M. S. *Dent Mater J* **2012**, *31*, 1054.
(15)   Wu, X.; Deng, F.; Wang, L.; Watts, D. C. *Microsc Res Tech* **2012**, *75*, 1176.
(16)   Wellner, P.; Mayer, W.; Hickel, R.; Reichl, F. X.; Durner, J. *Dent Mater* **2012**, *28*, 743.
(17)   Krifka, S.; Seidenader, C.; Hiller, K. A.; Schmalz, G.; Schweikl, H. *Clin Oral Investig* **2012**, *16*, 215.
(18)   Burke, F. J.; Crisp, R. J.; James, A.; Mackenzie, L.; Pal, A.; Sands, P.; Thompson, O.; Palin, W. M. *Dent Mater* **2011**, *27*, 622.
(19)   Goncalves, F. S.; Leal, C. D.; Bueno, A. C.; Freitas, A. B.; Moreira, A. N.; Magalhaes, C. S. *Am J Dent* **2013**, *26*, 93.
(20)   Gao, B. T.; Lin, H.; Zheng, G.; Xu, Y. X.; Yang, J. L. *Dent Mater J* **2012**, *31*, 76.
(21)   Crivello, J. V.; Mao, Z. B. *Chemistry of Materials* **1997**, *9*, 1554.
(22)   Yang, X.; Huang, W.; Yu, Y. *Journal of Applied Polymer Science* **2011**, *120*, 1216.
(23)   Watts, D. C.; Cash, A. J. *Dent Mater* **1991**, *7*, 281.

Mater. Res. Soc. Symp. Proc. Vol. 1621 © 2014 Materials Research Society
DOI: 10.1557/opl.2014.286

Development of hydroxyapatite-mediated synthesis of collagen-based copolymers for application as bio scaffolds in bone regeneration

Didarul Bhuiyan[1], John Middleton[1] and Rina Tannenbaum[2]

[1]Department of Biomedical Engineering, University of Alabama at Birmingham, Birmingham, AL 35924, USA.
[2]Department of Materials Science and Engineering, Program in Chemical and Molecular Engineering, Stony Brook University, Stony Brook, NY 11794, USA.

ABSTRACT

Hydroxyapatite (HAP) is a biocompatible bio-ceramic whose structure and composition is similar to bone. However, its lack of strength and toughness have seriously hampered its applications as a bone graft substitute material. Attempts have been made to overcome these mechanical properties deficiencies by combining HAP bioceramic material with absorbable polymers in order to improve its mechanical properties. However, poor interfacial bonding between the HAP and the polymers has limited the benefits of such biocomposite structures. At the other end of the biomaterials spectrum is collagen, which constitutes the most abundant proteins in the body and exhibits properties such as biodegradability, bioadsorbability with low antigenicity, high affinity to water, and the ability to interact with cells through integrin recognition. These favorable properties renders collagen as a natural candidate for the modification and compatibilization of the polymer-HAP biocomposite. In this study, we developed a novel approach to the synthesis of a potential bone graft material, where the HAP moiety acts not only as a bioceramic filler, but also constitutes the initiator surface that promotes the in-situ polymerization of the adsorbable polymer of choice. The synthesis of poly(D,L-lactide-co-glycolide) (PLGA) polymer was catalyzed by nano-hydroxyapatite (nHAP) particles and upon reaction completion, the biocomposite material was tethered with collagen. The synthesis was monitored by ^{1}H NMR and FTIR spectroscopies and the products after each step were characterized by thermal analysis to probe both thermal stability, morphological integrity and mechanical properties.

INTRODUCTION

Bone grafts can serve as scaffolds onto which natural bone can grow and regenerate and hence, can be used in the process of replacing deficient bone in order to repair fractures and defects caused by congenital disorders, traumatic injury, or resection of bone tumors.[1,2] Autogenous bone graft sources are widely used because they have a lower frequency of rejection by the immune system. However, in some cases, availability is limited and some complications, such as chronic pain, scarring, bleeding, and infection at the removal site, may occur. Allogeneic bone grafting is another option, but it carries with it increased probability of rejection.

Extensive ongoing research is devoted to the development of biocompatible scaffold materials that are made of synthetic or natural biomaterials and that would promote migration, proliferation and differentiation of bone cells for bone regeneration.[3-4] At present, there are no synthetic bone graft substitute materials available that have similar or superior biological and mechanical properties compared to bone. Hydroxyapatite (HAP) bio-ceramic is currently used in clinic as a bone graft substitute material. However, pure HAP bio-ceramic has low mechanical strength and weak fatigue resistance and therefore, is only suitable for repairing small fractures or bone defects that do not involve significant load bearing.[5] Researchers have attempted to overcome these limitations by compounding HAP with absorbable polymers, such as poly(D,L-lactide) (PLLA). However, weak interfacial bonding between the HAP and PLLA moieties gave rise to material incompatibility and filler-polymer phase separation. One approach to overcome this problem was to graft short chains of PLLA directly onto the surface of nanocrystalline hydroxyapatite and then blend up to 40 wt% of the PLLA-decorated particles into the PLLA matrix.[6] The result was an increase in ductility without compromising biocompatibility. Despite this encouraging approach, the fabrication of synthetic bone graft substitutes remains a very challenging endeavor and the various recent studies in this area show that a single phase system does not necessarily provide all the required materials features.

The use of the biodegradable poly(D,L-lactide-co-glycolide) (PLGA) polymer system, based on D,L-lactide and glycolide monomers, would be more attractive than PLLA for bone repair applications because its degradation rate can be adjusted by altering the ratio of lactic to glycolic acids to match the rate of bone formation.[5,7-9] To enhance the interfacial compatibilization between PLGA and nHAP, a biocompatible, absorbable component would be required. Such a component is collagen, the most abundant proteins in the body. In this study, we developed a novel approach to the synthesis of a potential bone graft material, where the HAP moiety acts not only as a bioceramic filler, but also constitutes the initiating surface that promotes the in-situ polymerization of the adsorbable polymer of choice. The synthesis of poly(D,L-lactide-co-glycolide) (PLGA) polymer was initiated by nano-hydroxyapatite (nHAP) particles and upon reaction completion, the biocomposite material was tethered with collagen. The synthesis was monitored by ^{1}H NMR and FTIR spectroscopies and the products after each step were characterized by thermal analysis to probe both thermal stability, morphological integrity and mechanical properties.

EXPERIMENTAL DETAILS

D,L-lactide and glycolide monomer were obtained from Ortec, Inc. (Easley, S.C.). Hydroxyapatite Nanopowder (<200 nm) was purchased from Sigma Aldrich. Lyophilized calf skin collagen type I (M_w 1,000,000 g/mol) was purchased from Elastin, Inc. (Owensville, Missouri). Previously dried 0.84 g nano-hydroxyapatite (nHAP) was mixed with 39.42 g D,L-lactide and 10.58 g glycolide monomers in a glass reactor placed in an oil bath at 150 °C. The molten monomer and nHAP mixture was sonicated with a sonicator probe for 5 min, following by the addition of 0.05 g stannous octoate catalyst. Samples were collected at during the polymerization process at different time intervals and analyzed by ^{1}H NMR spectroscopy. After 2 h of polymerization, 0.2 g succinic anhydride was added to the reactor. After 4.5 h of polymerization, the reactor was taken out of the oil bath and the nHAP-PLGA co-polymer was collected. 10 g of the co-polymer was dissolved in 200 mL dried methylene chloride and 0.39 g N-hydroxysuccinimide (NHS) and 0.09 g N,N'-dicyclohexylcarbodiimide (DCC) was added. The reaction was allowed to continue for 20 h under nitrogen atmosphere. The polymer solution was then dissolved in ethyl acetate and precipitated in anhydrous diethyl ether. The precipitated polymer was collected, dried and stored at 4 °C. 160 mg calf skin collagen type I was dissolved in 223.8 mL of 1 mM HCl in an ice bath and subsequently diluted with 300 mL of 50 mM phosphate buffer (pH 7.4). 0.96 g nHAP-PLGA-NHS polymer was dissolved in 26.6 mL anhydrous DMF and added drop wise into the collagen solution in the ice bath. The precipitated co-polymer sample was collected every 15 minutes for thermogravimetric analysis (TGA) experiments. After 3 h of reaction, the n-HAP-g-poly(D,L-lactide-co-glycolide)-

g-collagen block co-polymer was filtered, dried and stored at 4 °C. NMR samples were prepared by dissolving the co-polymers in CDCl₃, purchased from Fisher Scientific. ¹H NMR spectra were measured on a Bruker DPX-300 spectrometer at 300 MHz. The ¹H spectra for samples collected during the polymerization at various time intervals minutes were used to determine the ratio of monomer to polymer in the samples, and subsequently the kinetic behavior. The TGA experiments were conducted at a heating rate of 20 °C·min⁻¹ from room temperature to 600 °C in an unsealed platinum pan under nitrogen atmosphere and measured on a Du Pont Instruments 2950 Thermogravimetric Analyzer. Dynamic mechanical analysis was conducted on a TA Instruments RSA G2 instrument.

RESULTS AND DISCUSSION

The synthesis of the novel nano-hydroxyapatite (nHAP)-g-poly(D,L-lactide-*co*-glycolide)-g-collagen multi-component polymer was carried out in several steps. First, the polymerization of the poly(D,L-lactide-*co*-glycolide) (PLGA) polymer was initiated by the n-HAP nanoparticles by a ring-opening polymerization reaction. The progression of the polymerization reaction was monitored by ¹H NMR at various time intervals for two chemical shift ranges, 1.3–1.7 ppm corresponding to CH_3 groups and 4.6–5.3 ppm corresponding to CH_2 and CH groups. By integrating the area under each of the pertinent peaks, the ratio between D,L-lactide and glycolide monomers within the PLGA polymer could be determined, indicating the progress of the reaction. The fraction of the D,L-lactide and glycolide monomers in the PLGA polymer was followed as a function of reaction time, as shown in Figure 1a. The ratio of the two monomers in the PLGA polymer in the initial period of polymerization is about 50:50, indicating a random copolymerization process. As reaction progresses, the glycolide monomer becomes exhausted, and the remaining D,L-lactide monomer forms a block in the polymer, resulting in a polymer with mixed random and block architectures. At the end, after the purification and activation with NHS and DCC, the ratio of D,L-lactide and glycolide monomers in the polymer was 73:27, which is very close to the initial target composition of 75-25 in the PLGA polymer.

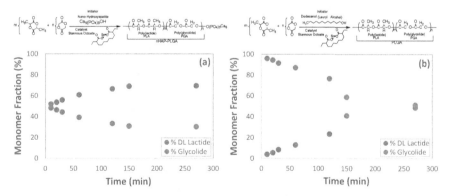

Figure 1: The fraction of the monomers in the polymer measured and calculated at various reaction times representing the uptake of these monomers in the growing polymer chain. (a) Reaction initiated by nHAP, (b) Reaction initiated by dodecanol.

When the nHAP-initiated reaction is compared to the same polymerization of D,L-lactide and glycolide to form PLGA that was initiated by dodecanol under otherwise identical conditions, a completely different reaction profile is observed, as shown in Figure 1b. In the initial stages of the reaction, the glycolide monomer is preferentially taken up into the polymer, despite its lower concentration in the reactant solution. Only after about 150 min. the uptake of monomers into the growing polymer chain become random. At the end of the reaction, the ratio of the two monomers in the polymer chain is about 50:50, which implies that a fraction of the D,L-lactide monomer has not reacted. As was the case with the nHAP-initiated reaction, the polymer formed is characterized by a mixture of random and block architectures.

In the second step, the PLGA co-polymer was activated prior to collagen attachment. The carboxyl end group obtained from succinic anhydride was reacted with N-hydroxysuccinimide (NHS) in the presence of N,N'-dicyclohexylcarbodiimide (DCC) as the cross-linking agent. In the third and final step, the activated PLGA co-polymer was attached to calf skin collagen type I in a hydrochloric acid/phosphate buffer solution and the precipitated co-polymer with the attached collagen was isolated. The number of PLGA chains was estimated by multiplying the number of moles of PLGA by N_A (Avogadro's number) and was found to correspond to 1×10^{21}. According to the nHAP manufacturer, the diameter of the nanoparticles was ≤ 200 nm. Considering the volume of the 200 nm nHAP particles, the surface area for each particle was calculated to be 1.25×10^5 nm^2. From the amount of nHAP used we estimated that the number of nHAP particles involved in the reaction was 6.39×10^{13}. Therefore, the total surface area of the nHAP particles was estimated as 8.03×10^{18} nm^2 and hence, the number of PLGA chains per nm^2 was calculated to be approximately 125. Using N_A, we have also estimated that there were 5.21×10^{18} collagen molecules involved in the reaction, which implies that on average, there is 1 collagen molecule attached to 192 PLGA chains. This may indicate that the collagen acts effectively as a cross-linking molecule, giving rise to a hydrogel-type material, as shown schematically in Figure 2.

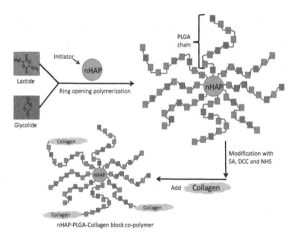

nHAP-PLGA-Collagen block co-polymer

Figure 2: The schematic description of chain organization in the nHAP-PLGA-collagen block co-polymer. The red squares in the PLGA chain represent glycolide molecules and blue squares represent D,L-lactide molecules. Calculations imply there are 125 chains of PLGA per nm^2 surface of nHAP and on average there is 1 collagen molecule attached for every 192 PLGA chains.

The thermal and mechanical properties of the biomaterial were then investigated by means of thermal analysis and dynamic mechanical testing. The thermal stability of the nHAP-PLGA-collagen co-polymer was measured by thermo-gravimetric analysis (TGA). The TGA decomposition profiles of samples collected at different stages during the collagen attachment process were compared to the decomposition profile of PLGA and collagen, and are shown in Figure 3. It is clear that tethering the collagen to the PLGA copolymer has had a noteworthy effect on the thermal stability of the hybrid material, beyond what would have been expected by assuming a simple composite effect. It is very likely the attachment of the collagen to the PLGA polymer has intimately altered the morphology of the copolymer, which in turn, has enhanced the thermal properties compared to the nHAP-PLGA copolymer and the collagen itself.

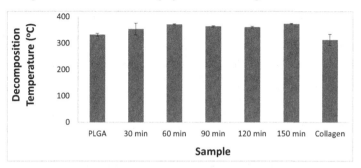

Figure 3: Average decomposition temperature for the co-polymer samples at different time intervals of collagen addition during synthesis process. The solid line represents the average decomposition temperature obtained by a linear fit of the data.

Dynamic mechanical analysis (DMA) was carried out and the moduli of the various biomaterials at various stages during the reaction are shown in Figure 4. It is interesting to note that the presence of the nHAP nanoparticles have a plasticizing effect on the PLGA polymer (as compared to PLGA synthesized with the dodecanol initiator), as shown in Figure 4b. Conversely, the presence of collagen has a profound effect on the magnitude of the modulus, as shown in Figure 4c. Given the calculated ratio between the collagen and the PLGA chains, it seems that the collagen acts as a cross-linking agent for the system, generating a stronger material.

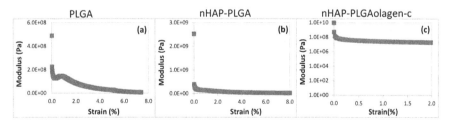

Figure 4: Moduli of the co-polymer biomaterial samples as measured by dynamic mechanical analysis instrument (DMA).

CONCLUSIONS

In the study, PLGA polymer was successfully grafted on the surface of nano particles of HAP. Chemical linkage was formed between nHAP particles and PLGA polymer and verified with ^{1}H NMR analysis. The nHAP grafted PLGA polymer was subsequently attached to collagen type I, and cross-linked co-polymer was obtained. TGA analysis shows that the samples behaved as a homogeneous material and the amount of impurity within the samples was negligible. TGA results reveal that the novel nHAP-PLGA-collagen co-polymer possesses higher thermal stability than the nHAP-PLGA polymer. Moreover, DMA analysis confirmed that the nHAP-PLGA-collagen system was stronger than PLGA most likely due to the cross-linking effect of the collagen.

REFERENCES

1. Branemark, P. I.; Worthington, P.; Grondahl, K. *Osseointegration and Autogenous Onlay Bone Grafts: Reconstruction of the Edentulous Atrophic Maxilla*; Quintessence Pub Co.: Chicago, IL, **2001**, p 4-8.

2. Beauchamp, D. R.; Evers, M.B.; Mattox, K. L.; Townsend, C. M.; Sabiston, D. C. *Sabiston Textbook of Surgery: The Biological Basis of Modern Surgical Practice,* 19th ed. W B Saunders Co.: London, **2012**, p 783-791.

3. Porjazoska, A.; Yilmaz, O.K.; Baysal, K.; Cvetkovska, M.; Sirvanci, S.; Ercan, F.; Baysal, B. M. Synthesis and Characterization of poly(ethylene glycol)-poly(D,L-lactide-co-glycolide) poly(ethylene glycol) tri-block co-polymers modified with collagen: a model surface suitable for cell interaction. *J. Biomater. Sci. Polymer Ed.* **2006**, 17(3), 323-340.

4. Liao, S.; Wang, W.; Uo, M.; Ohkawa, S.; Akasaka, T.; Tamura, K.; Cui, F.; Watari, F. A three-layered nano-carbonated hydroxyapatite/collagen/PLGA composite membrane for guided tissue regeneration. *Biomaterials*, **2005**, 26, 7564-7571.

5. Zhang, P.; Hong, Z.; Yu, T.; Chen, X.; Jing, X. In Vivo mineralization and osteogenesis of nanocomposite scaffold of poly(lactide-co-glycolide) and hydroxyapatite surface-grafted with poly(L-lactide). *Biomaterials*, **2009**, 30, 58-70.

6. Hong, Z.; Qiu, X.; Sun, J.; Deng, M.; Chen, X.; Jing, X. Grafting polymerization of L-lactide on the surface of hydroxyapatite nano-crystals. *Polymer*, **2004**, 45, 6699-6706.

7. Furukawa, T.; Matsusue, Y.; Yasunaga, T.; Shikinami, Y.; Okuno, M.; Nakamura, T. Biodegradation behavior of ultra-high-strength hydroxyapatite/poly(L-lactide) composite rods for internal fixation of bone fractures. *Biomaterials*, **2000**, 21, 889-898.

8. Hasegawa, S.; Ishii, S.; Tamura, J.; Furukawa, T.; Neo, M.; Matsusue, Y.; Shikinami, Y.; Okuno, M.; Nakamura, T. A 5-7 year in vivo study of high-strength hydroxyapatite/poly(L-lactide) composite rods for the internal fixation of bone fractures. *Biomaterials*, **2006**, 27, 1327-1332.

9. Shikinami, Y.; Okuno, M.; Bioresorbable devices made of forged composites of hydroxyapatite (HA) particles and Poly-L-lactide (PLLA): Part I. Basic characteristics. *Biomaterials*, **1999**, 20, 859-877.

Mater. Res. Soc. Symp. Proc. Vol. 1621 © 2014 Materials Research Society
DOI: 10.1557/opl.2014.67

Size Also Matters in Biodegradable Composite Microfiber Reinforced by Chitosan Nanofibers

Elisabete D. Pinho[1,2], Albino Martins[1,2], José V. Araújo[1,2], Rui L. Reis[1,2] and Nuno M. Neves[1,2]

[1] 3B's Research Group – Biomaterials, Biodegradables and Biomimetics. Department of Polymer Engineering, University of Minho; Headquarters of the European Institute of Excellence on Tissue Engineering and Regenerative Medicine. AvePark, Zona Industrial da Gandra, S. Cláudio do Barco 4806-909 Caldas das Taipas, Guimarães, Portugal.
[2] ICVS/3B's, PT Government Associate Laboratory, Braga/Guimarães, Portugal.

ABSTRACT

Pioneer works on nanocomposites were focused in carbon nanofibers or nanotubes dispersed in epoxy matrix, a viscous liquid facilitating the compounding stage. The interest in developing new composites aimed for biomedical applications led us to design new nanocomposites based in biodegradable polymers with demonstrated biological performance.

We report herein the development of micro-nano composites by extruding poly(butylene succinate) (PBS) microfibers with two different diameters, 200 and 500 μm, reinforced with electrospun chitosan nanofibers. Analysis of the microfibers showed high levels of alignment of the reinforcing phase and excellent distribution of the nanofibers in the composite. Its geometry facilitates the development of orthotropy, maximizing the reinforcement in the axial fiber main axis.

The biodegradable microfiber composites show an outstanding improvement of mechanical properties and of the kinetics of biodegradation, with very small fractions (0.05 and 0.1 wt.%) of electrospun chitosan nanofibers reinforcement. The high surface area-to-volume ratio of electrospun nanofibers combined with the increased water uptake capability of chitosan justify the accelerated kinetics of biodegradation of the composite as compared with the unfilled synthetic polymer.

INTRODUCTION

Nanocomposites are a class of advanced materials consisting in a reinforcing phase with some dimensions in the nanometer size range dispersed in a continuous matrix. Usually the nanometer dimension range is defined between 1 and 100 nm. This requirement is not consensually accepted, depending on the characteristics of the nanoentities of interest and on the field of application [1]. The nanocomposites may exhibit enhanced properties such as the elastic modulus, strength, heat resistance, barrier properties, transparency or biodegradability. The superior properties of nanocomposites have potential interest for many applications in electronic, automotive or biomedical industries [2].

Fibrous composites allow easier structuring of reinforcements being more effective than bulk materials. Its geometry facilitates the development of orthotropy, maximizing the reinforcement in the fiber main axis. The production of composite fibers by melt processing using extrusion is a very well established technology [3,4]. The melt flow dynamics in the extrusion of fibers facilitate the alignment of reinforcements within the composite and the reinforcing efficiency.

The maximum reinforcement efficiency is obtained when the nanofibers are uniformly dispersed and individually coated with the continuous matrix of the composite [3]. A good fiber dispersion facilitates the load transfer and development of uniform stress distribution and minimizes the probability of development of structural defects or voids acting as stress-concentration points. The alignment also allows maximizing strength and stiffness in particular directions [5]. Composites intended for biomedical applications such as wound care, drug delivery, medical devices or scaffolds for tissue engineering have the critical requirement of the biocompatibility. The biodegradability is also very important if those materials are developed for temporary devices such as resorbable sutures or tissue engineering scaffolds [4,6].

The pioneer work on nanofiber composites was focused in carbon nanofibers or nanotubes [3]. Those nano-reinforced composites may be produced by polymer processing technology using solvents and surfactants to avoid nanofiber agglomeration. Electrospun polymer nanofiber meshes were explored as composite reinforcements in a small number of studies [4,7]. Those reports were mainly centered in the reinforcing effect of electrospun nanofibers dispersed in an epoxy matrix [7].

Electrospinning is an efficient technique for the production of nanofibers with diameters in the submicrometer range [8,9]. The increased surface area-to-volume ratio as a consequence of the submicron diameter is one of the most interesting properties of electrospun fibers [10,11]. The nanoscale size of the biodegradable fibers may also offer advantages in terms of inducing specific degradation kinetics. In this study we successfully produced and stabilized electrospun nanofibers from chitosan. Chitosan is an alkaline deacetylated derivative of chitin that may be obtained from the shells of arthropods or cephalopods. This material in combination with various biodegradable polyesters has already shown a good combination of properties for different biomedical applications [12-15]. The poly(butylene succinate) (PBS) matrix showed previously stronger biological performance in combination with chitosan, being thus selected for the present study. The biodegradable aliphatic polyester PBS, presenting a hydrophobic character, shows slow degradation kinetics. Chitosan is rich in polar groups (-OH and $-NH_2$) and is very hydrophilic. Its presence as reinforcement in the microfiber composite increases the final hydrophilicity of the composite, typically being determined by the wettability of the continuous phase [16]. The high surface area-to-volume ratio combined with the increase in the water uptake capability of polymer reinforced composite fiber is intended to accelerate the kinetics of biodegradation of the composite.

Fibers of poly(butylene succinate) (PBS) with and without the nanofiber reinforcement were produced in two diameters (200 and 500 µm) to evaluate the efficiency of reinforcement in those two fiber dimensions. The effect of the chitosan nanofiber meshes (Cht NFM) reinforcements concentration was evaluated by compounding the composites with 0.05% and 0.1% by weight. The analysis of the composite fibers included the study of there morphology, mechanical properties and the kinetics of degradation, after various periods of immersion in an isotonic saline solution. We also characterized the thermal properties of the composite by thermogravimetrical analysis and calorimetry, and the composites chemical composition by Attenuated Total Reflection-Fourier Transform InfraRed (ATR-FTIR).

EXPERIMENTAL DETAILS

Materials

Poly(butylene succinate) (PBS) Bionolle^{TM} 1050 was supplied by Showa Highpolymer Co. Ltd., Tokyo, Japan. Trifluoroacetic acid (TFA) was obtained from Sigma and Dichloromethane (DCM) was supplied by Aldrich. Ammonia 7N solution in methyl alcohol was purchased from Aldrich and Methanol was obtained from Fluka.

Production of Chitosan Nanofiber Meshes (Cht NFM)

The electrospinning processing and the subsequent neutralization process were conduct according to the protocol described elsewhere [4]. Briefly, a polymeric solution of 4% (w/v) chitosan (Mv=417kDa; DD=88%) was prepared using a solvent mixture of TFA and DCM (7:3 ratio). Electrospinning of this polymeric solution was achieved by establishing a electric field of 16-18 kV, a needle-to-ground collector distance of 12-14 cm and a flow rate of 0.6-0.8 mL/h. Neutralization of Cht NFM was performed by immersing them in a solution containing concentrated ammonium and methanol.

Production of Microfibers by Melt Extrusion

Three different types of compositions were produced in this study: i) PBS fiber; ii) Fiber composed by PBS (99.95% by weight) reinforced with 0.05% by weight of Cht NFM; iii) Fiber composed by PBS (99.9% by weight) reinforced with 0.1% by weight of Cht NFM. The processing conditions were optimised to obtain a stable extrusion processing. The processing conditions were kept constant for all compositions to maintain the same thermal history in all compositions. The melt temperature was set at 115 ºC and the screw rotation speed at 40 r.p.m. The die diameters used were either 0.2 or 0.5 mm for each composition.

Morphological Analysis

The morphology of the microfiber composites (with and without nanofiber mesh reinforcements) was analysed by scanning electron microscopy (SEM) (model S360, Leica Cambridge) after being sputter coated with gold. To analyse the morphology, distribution and structure of the chitosan nanofibers inside of the composite microfibers, a cryogenic fracture of the fiber or a partial dissolution of the polyester matrix with dichloromethane was performed.

Surface Chemistry Characterization

The chemical characterization of the surfaces was firstly performed by Attenuated Total Reflection-Fourier Transform InfraRed (ATR-FTIR). The FTIR spectra was recorded on an IRPrestige 21 FTIR spectrophotometer (Shimadzu, Japan) with a resolution of 4 cm^{-1} and averaged over 36 scans.

Differential scanning calorimetry (DSC)

The measurements were carried out using a Perkin Elmer DSC-7 instrument under a dynamic nitrogen atmosphere. The samples were heated to 30 °C to 150 °C by 10 °C/min. The mass of the samples was around 5 mg. The calculations were making by Avrami model. The degree of crystallinity of the PBS component was calculated from the known theoretical value of ΔH_f can be compared with 110.3 J/g, corresponding to 100% crystalline PBS calculated on the basis of the group contribution method proposed by Van Krevelen [17].

Thermogravimetric analysis (TGA)

A thermogravimetric analyzer (Q500, TA Instruments, New Castle-Delaware, USA) was used to investigate the effect of Cht NFM processing on the polymer. PBS fibers (without reinforcement) and PBS fibers reinforced by Cht NFM were analyzed in closed Platinum cups in a temperature range of 30-1000 °C. The heating rate established was 10 °C/min. All the experiments were carried out in a nitrogen atmosphere (flow rate 60 cm^3/min).

Mechanical Tests

The reinforced composite fibers were cut in 10 mm long fiber specimens with diameter of approximately 200 and 500 μm. The tensile tests were performed at room temperature in a Universal Mechanical Test Machine (Instron 4505) using a load cell of 1 KN and a crosshead speed of 5 mm/min. A minimum of six specimens were tested in each sample (the values reported are the average of those results).

Swelling and Degradation-Related Tests

The hydration degree and degradation behaviour of the particulate microfiber compositions were assessed over a period of 30 days. Five specimens of each sample (previously weighed) were immersed in an isotonic saline solution (ISS: 0.154M NaCl aqueous solution, pH = 7.4) during 1, 3, 7, 14 and 30 days.

RESULTS & DISCUSSION

The morphology of the microfiber composites (with and without nanofiber mesh reinforcements) was analyzed by scanning electron microscopy (SEM) and are showed in Figure 1 a) and b). The fiber, shown in Figure 1 a), was produced with a die diameter of 500 μm (resulting in a fiber diameter of approximately 460 μm). In Figure 1 b) it is shown a fiber produced with a die diameter of 200 μm (fiber diameter of approximately 210 μm). The fibers are characterized by a regular and smooth surface confirming the effectiveness of the extrusion processing.

The chitosan nanofiber mesh obtained by electrospinning is soluble in water and is not structurally stable. Its use in biomedical applications implies a neutralization process to chemically stabilize the chitosan but preserving the morphology of the nanofiber mesh. In a previous work we reported that electrospun Cht NFM, after the neutralizing alkaline treatment,

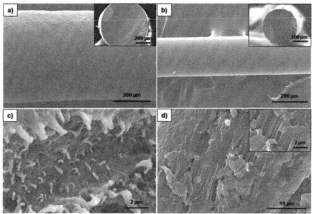

Figure 1. SEM micrographs of: **a)** fiber with 460 μm of diameter (fiber*500*), **b)** fiber with 210 μm of diameter (fiber*200*), **c)** cross-section of PBS fiber reinforced by Cht NFM and **d)** longitudinal section.

successfully maintained the electrospun nanofiber morphology [4]. The fibers have diameters ranging from 65 nm to 6 μm, exhibiting a regular and smooth surface topography without the presence of beads which is commonly referred as a problem of electrospun nanofiber meshes [9]. The mesh-like structure has interconnected pores and a non-woven, random distribution. Those observations confirm the efficacy of the optimized conditions used both in the electrospinning process and in the subsequent neutralization treatment.

As expected, the Cht nanofibers are present underneath the fiber surface in the inner core of the fibers (being only exposed after cryogenic fracture of the fiber sample). The reinforced composite fiber combines the micrometer size of the extruded fiber with the nanoscale of the reinforcement. The biodegradable PBS matrix reinforced by Cht NFM shows the nanofiber mesh highly aligned after the processing. This alignment is not present before the extrusion process and is probably generated by the hydrodynamics of the flowing melt during the extrusion process [18]. During polymer melt extrusion a stress field develops caused by the viscous nature of the melt. The stresses acting in the melt are predominantly shear stresses that are known to cause the alignment of fibers in the main flow direction. This effect can be observed both in cross section (Figure 1 c) and also in longitudinal section (Figure 1 d) micrographs. The micrographs show the alignment of the nanofibers with the main axis of the fiber and the intimate coverage of the Cht NFM by the continuous phase provided by the extrusion process. The distribution of the nanofibers and the adhesion matrix/nanofibers (Figure 1 c and d) shows an excellent interface without visible voids or defects. The high alignment of the Cht NFM in the composite suggests that significant anisotropy of mechanical properties may exist [3-5].

The surface chemistry of the fiber composites, the PBS fibers and the chitosan nanofiber meshes were all analyzed by FTIR-ATR. The spectra of the different fiber materials and composites are shown in Figure 2 a). The characteristic absorption bands were assigned to the

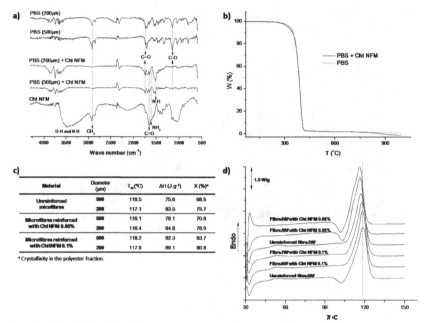

Figure 2. Chemical and thermal analysis: **a)** FTIR-ATR for the unreinforced and reinforced PBS fibers by Cht NFM; **b)** Representative thermogravimetric curves of the processing fibers; **c)** Melting temperature and heat of fusion of the different samples analyzed by DSC, and the corresponding crystallinity degree within the synthetic polymer component; **d)** DSC thermograms for unreinforced fibers and fibers reinforced by Cht NFM.

components of the reinforced composite fiber. The stretching bands corresponding to O-H and N-H (3480-3080 cm^{-1}), CH$_2$ (2960-2560 cm^{-1}), C=O (1648 cm^{-1}), NH$_2$ (1612 cm^{-1}) and N-H (1509 cm^{-1}) are all assigned to chitosan. The stretching bands corresponding to C=O (1741 cm^{-1}), C-O (1140 cm^{-1}) and C-H (1147-1263 cm^{-1}) are assigned to PBS [16]. As was expected, the spectrum of the reinforced composite fiber presented the main chemical groups of the Cht NFM. In some cases the lower intensity bands of chitosan were overlapped by the much sharper and higher intensity bands of PBS and, then, are more difficult to be detected in the reinforced composite fibers.

Figure 2 b) shows the TGA curves for the materials subjected to extrusion processing. TGA was used in this study to observe the effect of Cht NFM on the thermogravimetrical properties of the composite. It is important to verify if the processing temperatures used in extrusion cause any significant thermogravimetrical variation. The small mass loss observed at the extrusion temperature is probably due to loss of moisture or minor volatile reagents present in the material as processing additives. The presence of 0.1% of Cht NFM in the fiber composite does not affect the pattern of thermal degradation. TGA data shows that the thermal degradation initiates at approximately 230 °C. Most of the material undergoes volatilization at 416 °C. Both profiles showed similar thermal degradation. The infrared spectroscopy results together with the TGA analysis allow confirming the absence of thermal degradation of chitosan in the produced reinforced composite fibers.

The results from differential scanning calorimetry (DSC) experiments for unreinforced fibers and fibers reinforced by Cht NFM are shown in Figure 2 d). The melting of the synthetic phase was clearly detected in all the fibers (endothermic peak). Figure 2 c) summarizes the thermal characteristics of the fibers. It is noted the increase of the ΔH with increasing Cht NFM content in the PBS fibers. The degree of crystallinity of the PBS component was calculated from the estimated theoretical value of ΔH for 100% crystalline PBS, having a value 110.3 J/g. This value was calculated on the basis of the group contribution method proposed by Van Krevelen [17]. PBS-based materials show some tendency for the depression of Tm with the introduction of the Cht NFM reinforcement, except for the fiber*500* and fiber*200* reinforced with 0.1% of Cht NFM. This trend may be caused by the chitosan fibers acting as a nucleating agent and lowering the size of the crystalline structures. This phenomenon could also explain the increase in the level of crystallinity observed in the reinforced composite fibers. This possible effect of Cht NFM over the nucleation and development of crystallinity of the PBS matrix may also influence the mechanical properties of the composite, eventually by creating a strong interface between the matrix and the reinforcement [3, 5].

The tensile modulus, tensile stress and tensile strain are all shown in Table I. A value of 31 MPa was obtained for the tensile yield stress (strength) of the unreinforced fiber*500* and tensile strain of 12%. The unreinforced fiber*200* presents much higher values of both the tensile stress and strain, 75 MPa and 46% respectively. The fiber*500* without reinforcement has tensile modulus of 329 MPa and the fiber*200* shows much higher tensile modulus (950 MPa). This result may be explained by the increase in molecular orientation experienced by PBS when extruded into a thinner fiber. The mechanical properties of the fibers*200* can be also interpreted together with DSC results (Figure 2 c). DSC results show that the unreinforced fiber*200* presents a higher crystallinity that directly influences the mechanical properties. A higher degree of crystallization corresponds typically to higher modulus and tensile strength. However, the differences in crystallinity observed are not sufficient to explain the threefold increase in tensile modulus.

Despite the modest fraction of Cht NFM reinforcement, the reinforced fibers show both tensile modulus and tensile stress significantly enhanced. The maximum tensile strength was obtained for fiber*200* reinforced with 0.05% of Cht NFM having a value of 154 MPa. This value is much higher than that obtained for fiber*500* with 0.1% of Cht NFM. The tensile strain properties of fibers reinforced by Cht NFM shows a reduction of the ultimate tensile strain with increasing Cht NFM content. This result is also an indication of a strong reinforcement of the fibers by the presence of the Cht NFM, having associated a typical decrease in tensile strain [19]. The PBS fiber reinforced by Cht NFM showed a maximum increase in the tensile modulus in the case of fiber*200* with 0.1% of Cht NFM. A value of the modulus of 2017 MPa was obtained, representing the double of that obtained for unreinforced fiber*200* (949.8 MPa). The composite fiber*500* with 0.05% of Cht NFM showed a tensile modulus of 553.2 MPa, 70% larger than the unreinforced fiber*500* (329.3 MPa). The reinforcement by Cht NFM is much less effective in fiber*500* than in fiber*200*. The reinforcement resulting from the orientation along the main axis of the fiber micro/nanocomposites probably is probably responsible for those differences.

Table I. Tensile properties of fiber*500* and fiber*200* reinforced or not with Cht NFM, obtained from the tensile curves, before and after being subjected to swelling and degradation tests.

Materials	Immersion Time [days]	Diameter [μm]	Tensile Strength [MPa]	Tensile Modulus [MPa]	Tensile Strain [%]
Unreinforced microfibers	0	500	30.6 ± 7.2	329.3 ± 36.9	11.6 ± 2.3
		200	74.5 ± 11.1	949.8 ± 197.3	46.0 ± 10.2
	1	500	19.3 ± 6.3	322.1 ± 87.6	8.8 ± 1.2
		200	46.9 ± 5.9	597.8 ± 62.2	41.4 ± 16.7
	3	500	18.2 ± 4.9	308.1 ± 8.4	8.0 ± 1.5
		200	51.1 ± 12.8	594.3 ± 73.6	14.3 ± 4.3
	7	500	17.7 ± 6.4	302.8 ± 64.3	7.3 ± 0.8
		200	35.6 ± 10.3	654.7 ± 128.6	15.9 ± 11.6
	14	500	11.2 ± 3.4	140.8 ± 54.6	8.8 ± 1.5
		200	28.3 ± 7.2	540.2 ± 62.9	15.0 ± 7.5
	30	500	5.6 ± 1.6	129.8 ± 61.9	4.6 ± 1.0
		200	7.1 ± 1.5	425.0 ± 136.0	5.8 ± 2.5
Microfibers reinforced with Cht NFM 0.05%	0	500	35.4 ± 6.4	553.2 ± 48.4	10.7 ± 0.9
		200	154.0 ± 34.2	1783.5 ± 238.5	29.2 ± 10.2
	1	500	32.6 ± 9.5	513.6 ± 77.3	11.2 ± 2.3
		200	104.3 ± 12.8	1085.5 ± 330.6	32.5 ± 21.7
	3	500	25.5 ± 2.6	492.7 ± 55.1	7.6 ± 2.5
		200	55.5 ± 11.8	858.7 ± 137.0	12.8 ± 4.4
	7	500	24.9 ± 6.9	475.9 ± 34.3	8.9 ± 3.5
		200	42.7 ± 10.2	1127.7 ± 10.2	7.5 ± 1.8
	14	500	18.3 ± 1.2	446.3 ± 113.3	7.4 ± 1.9
		200	19.6 ± 7.7	736.4 ± 129.0	8.1 ± 6.4
	30	500	11.7 ± 2.8	395.9 ± 33.7	4.1 ± 1.1
		200	14.0 ± 4.5	629.6 ± 134.0	9.2 ± 5.9
Microfibers reinforced with Cht NFM 0.1%	0	500	39.0 ± 8.3	543.4 ± 69.0	17.3 ± 4.5
		200	136.7 ± 23.9	2016.7 ± 131.2	19.3 ± 10.7
	1	500	28.7 ± 2.5	572.6 ± 43.1	9.8 ± 2.9
		200	72.1 ± 22.7	1046.1 ± 187.8	19.9 ± 8.9
	3	500	15.3 ± 6.3	459.2 ± 39.0	7.2 ± 3.8
		200	38.1 ± 8.2	716.7 ± 170.8	11.3 ± 5.1
	7	500	19.3 ± 4.0	491.7 ± 90.9	6.7 ± 0.7
		200	33.8 ± 14.2	931.4 ± 158.4	13.9 ± 8.8
	14	500	8.7 ± 4.1	437.3 ± 84.7	6.4 ± 2.9
		200	29.9 ± 10.9	802.4 ± 167.2	10.0 ± 8.1
	30	500	7.5 ± 3.0	326.5 ± 87.6	7.5 ± 2.5
		200	15.8 ± 4.9	658.0 ± 107.9	4.5 ± 1.4

Continuous and highly aligned reinforcements in composites are more effective in achieving higher mechanical properties and also higher anisotropy (or orthotropy) [20]. Additionally, the nanofibers being longer and well oriented provide more surface area to develop a good interface with the matrix. As a result, not only the stiffness but also the tensile strength of the fibers reinforced by Cht NFM may be increased (Table I). The interface properties are critical to ensure the load transfer process and increase the reinforcing efficiency. The submicron diameters of the Cht NFM also help in avoiding the formation of defects that could limit the tensile strength.

Biomaterials developed to operate in immersion conditions should be evaluated in terms of swelling and hydrophilic/hydrophobic character. The rate of water uptake is related to various physical and chemical properties such as hydrophilicity, crystallinity or surface area [21]. Figure 3 a) and b) show the water absorption profiles of Cht NFM, fiber500 and fiber200 with and without reinforcement. Statistically significant differences in water uptake capacity are observed depending on the composition. The different diameters of the fibers also influence the water absorption being explained by the diffusion length. The fiber500 have maximum water uptake of 2.4% and the fiber200 have 5.4% of water uptake capacity. The smaller fiber diameter has also higher surface area per unit of volume, which further increments the water uptake ability. Figure 3 a) shows that the fiber500 and the fiber500 with 0.1% of Cht NFM reached a maximum of 8% in water uptake at day 14. In Figure 3 b) the fiber200 reinforced with 0.1% of Cht NFM increased the water uptake to 13.3%, reaching also a maximum at day 14. The low amount of Cht NFM used in the composites shows an outstanding increment in water absorption. The results are probably related with the high surface area of the nanofibers, facilitating the intake of water, and the properties of chitosan, a highly hydrophilic natural polymer (Cht NFM water uptake is 300%). The limit value of the water uptake may be regarded as a result of two

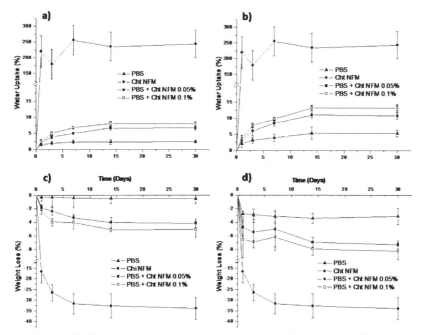

Figure 3. a) and **b)** Water uptake percentage along experimental time course for fiber500 and fiber200, respectively, with or without reinforcement and Cht NFM; **c)** and **d)** represents the weight loss percentage along different degradation times of fibers reinforced by Cht NFM, chitosan nanofiber meshes and unreinforced fibers.

complementary phenomena: the intrinsic water uptake capability of the material; and the degradation history. Those phenomena are somewhat inter-related, because the water absorption usually accelerates the degradation process, particularly in materials that are sensitive and undergo hydrolysis [22].

Figure 3 c) and d) show the weight loss results of the fibers with and without Cht NFM reinforcements. The weight loss of the unreinforced fiber*500* and fiber*200* reached a maximum value of 0.5% and 3.4%, respectively, after being subjected to 30 days of immersion. This shows that the degradation is still slow in the composite, but faster than that of PBS.

The mechanical properties of the fibers after the degradation tests decrease, as expected (Table I). The fibers reinforced by Cht NFM show decrease of modulus with time of immersion. The tensile modulus of fiber*500* reinforced with 0.05% of Cht NFM after 30 days of immersion was higher than that of unreinforced fibers before being subject to immersion tests (the initial tensile modulus). This result (Table I) may be also considered a remarkable property of these reinforced composite fibers.

Composites reinforced with 0.1% of Cht NFM (fiber*500* and fiber*200*) do not show dramatic differences except in the kinetics of degradation of fiber*200* probably caused by the larger surface area facilitating the diffusion. The variation of tensile strength during the periods of immersion was not as significant as the variation in the tensile modulus. The fibers reinforced by Cht NFM show comparable values of the ultimate tensile strain when compared with the unreinforced ones after 30 days of immersion. The mechanical properties clearly increased by the composition with Cht NFM, as well as the degradation kinetics previously reported.

CONCLUSIONS

Novel micro-nano composites were developed by producing biodegradable polymeric fibers reinforced by electrospun polysaccharide nanofibers. The nanofibers, although being initially obtained in randomly aligned meshes, have a well-aligned morphology inside the composite fibers. The fibers reinforced by nanofibers show a considerable alignment of the Cht NFM along its longitudinal main axis. This alignment is caused by the dynamics of the flowing melt during extrusion and by the stress fields. The Cht NFM reinforcement in the fibers also increased the water uptake. The weight loss was increased, indicating that the kinetics of biodegradation was significantly accelerated by the presence of the nanofiber reinforcement. Also the diameter causes variation in the kinetics of biodegradation being faster in the thinner fibers.

The elastic modulus of the fiber composite was significantly enhanced, even with very modest contents of nanofiber reinforcements. The tensile strength was marginally improved and the tensile strain is reduced considerably as is typical for fiber-reinforced composites. Thinner fibers present the same trends but higher efficiency of reinforcement. The increase of the crystallinity with the introduction of Cht NFM in the fibers also positively contributes for enhancing the mechanical properties of the fiber composites. The new nanofiber composite shows an outstanding improvement of mechanical properties and kinetics of biodegradation with very modest fractions of reinforcement and represents a new strategy to design materials with tailored properties and enhanced performance for many applications.

ACKNOWLEDGMENTS

This work was partially supported by the OsteoGraphy (PTDC/EME-MFE/2008) and MaxBone (PTDC/SAU-ENB/115179/2009) projects, financed by the Portuguese Foundation for Science and Technology (FCT).

REFERENCES

1. E. Stodolak, C. Paluszkiewicz, M. Bogun and M. Blazewicz, *J. Mol. Struc.*, **924–926**, 208 (2009).
2. Q. Y. Soundararajah, B. S. B. Karunaratne and R. M. G. Rajapakse, *Mater. Chem. Phys.* **113**, 850 (2009).
3. J. N. Coleman, U. Khan and Y. K. Gun'ko, *Adv. Mater.* **18**, 689 (2006).
4. E. D. Pinho, A. Martins, J. V. Araújo, R. L. Reis and N. M. Neves, *Acta Biomater.* **5**, 1104 (2009).
5. L.-Q. Liu, D. Tasis, M. Prato and H. D. Wagner, *Adv. Mater.* **19**, 1228 (2007).
6. L. Y. Yeo and J. R. Friend, *J. Exp. Nanosci.* **1**, 177 (2006).
7. Z.-M. Huang, Y.-Z. Zhang, M. Kotaki and S. Ramakrishna, *Compos. Sci. Technol.* **63**, 2223 (2003).
8. J. Venugopal, P. Vadgama, T. S. S. Kumar and S. Ramakrishna, *Nanotechnology* **18**, 511 (2007).
9. A. Martins, R. L. Reis and N. M. Neves, *Int. Mater. Rev.* **53**, 257 (2008).
10. T. Lin, H. Wang and X. Wang. *Adv. Mater.* **17**, 2699 (2005).
11. A. Martins, E. D. Pinho, S. Faria, I. Pashkuleva, A. P. Marques, R. L. Reis and N. M. Neves, *Small* **5**, 1195 (2009).
12. V. M. Correlo, L. F. Boesel, M. Bhattacharya, J. F. Mano, N. M. Neves and R. L. Reis, *Macrom. Mater. Eng.* **290**, 1157 (2005).
13. V. M. Correlo, E. D. Pinho, I. Pashkuleva, M. Bhattacharya, N. M. Neves and R. L. Reis, *Macromol. Biosci.* **7**, 354 (2007).
14. A. R. Costa-Pinto, A. J. Salgado, V. M. Correlo, P. C. Sol, M. Bhattacharya, P. Charbord, R. L. Reis and N. M. Neves, *Tissue Eng. Part A* **14**, 1049 (2008).
15. J. T. Oliveira, V. M. Correlo, P. C. Sol, A. R. Costa-Pinto, P. B. Malafaya, A. J. Salgado, M. Bhattacharya, P. Charbord, N. M. Neves and R. L. Reis, *Tissue Eng. Part A* **14**, 1651 (2008).
16. D. F. Coutinho, I. Pashkuleva, C. M. Alves, A. P. Marques, N. M. Neves and R. L. Reis, *Biomacromolecules* **9**,1139 (2008).
17. D. W. Van Krevelen *Properties of polymers*. 3rd Ed. Amsterdam. Elsevier, 1990.
18. N. M. Neves, G. Isdell, A. S. Pouzada and P. C. Powell, *Polym. Compos.* **19**, 640 (1998).
19. H. Mahfuz, A. Adnan, V. K. Rangari, S. Jeelani and B. Z. Jang, *Composites: Part A* **35**, 519 (2004).
20. W. Krause, F. Henning, S. Tröster, O. Geiger and P. Eyerer, *J. Thermoplas. Compo. Mater.* **16**, 289 (2003).
21. V. M. Correlo, L. F. Boesel, M. Bhattacharya, J. F. Mano, N. M. Neves and R. L. Reis, *Mater. Sci. Eng. A* **403**, 57 (2005).
22. M. P. Pavlov, J. F. Mano, N. M. Neves and R. L. Reis, *Macromol. Biosci.* **4**, 776 (2004).

Mater. Res. Soc. Symp. Proc. Vol. 1621 © 2014 Materials Research Society
DOI: 10.1557/opl.2014.287

Proliferation and Osteogenic Differentiation of Mesenchymal Stem Cells on Biodegradable Calcium-deficient Hydroxyapatite Tubular Bacterial Cellulose Composites

Pelagie Favi[1], Madhu Dhar[2], Nancy Neilsen[3] and Roberto Benson[1]
[1]Materials Science and Engineering, University of Tennessee, Knoxville, TN 37996
[2]Large Animal Clinical Sciences, University of Tennessee, Knoxville, TN 37996
[3]Biomedical and Diagnostic Sciences. University of Tennessee, Knoxville, TN 37996

ABSTRACT

Advanced biomaterials that mimic the structure and function of native tissues and permit stem cells to adhere and differentiate is of paramount importance in the development of stem cell therapies for bone defects. Successful bone repair approaches may include an osteoconductive scaffold that permits excellent cell adhesion and proliferation, and cells with an osteogenic potential. The objective of this study was to evaluate the cell proliferation, viability and osteocyte differentiation of equine-derived bone marrow mesenchymal stem cells (EqMSCs) when seeded onto biocompatible and biodegradable calcium-deficient hydroxyapatite (CdHA) tubular-shaped bacterial cellulose scaffolds (BC-TS) of various sizes. The biocompatible gel-like BC-TS was synthesized using the bacterium *Gluconacetobacter sucrofermentans* under static culture in oxygen-permeable silicone tubes. The BC-TS scaffolds were modified using a periodate oxidation to yield biodegradable scaffolds. Additionally, CdHA was deposited in the scaffolds to mimic native bone tissues. The morphological properties of the resulting BC-TS and its composites were characterized using scanning electron microscopy. The ability of the BC-TS and its composites to support and maintain EqMSCs growth, proliferation and osteogenic differentiation *in vitro* was also assessed. BC-TS and its composites exhibited aligned nanofibril structures. MTS assay demonstrated increasing proliferation and viability with time (days 1, 2 and 3). Cell-scaffold constructs were cultured for 8 days under osteogenic conditions and the resulting osteocytes were positive for alizarin red. In summary, biocompatible and biodegradable CdHA BC-TS composites support the proliferation, viability and osteogenic differentiation of EqMSCs cultured onto its surface *in vitro*, allowing for future potential use for tissue engineering therapies.

INTRODUCTION

The ideal scaffold to treat damaged tissue resulting from traumatic injuries and diseases remains a challenge for tissue engineering of bone. Current clinical procedures to repair this tissue include the use of autographs and allographs [1]. However, autographs may be unavailable for use and allographs are often diseased or lack immune compatibility [2]. The current clinical treatment for bone tissue damages and diseases stresses the need for an alternative replacement therapy. Thus, developing advanced biomaterials that can mimic the chemical and structural nature of native bone tissue, permit stem cells to adhere and differentiate to regenerate the lost bone function is important in the development of stem cell therapies for bone defects.

Native bone tissue is a three-dimensional (3D) hierarchical tubular structure consisting of extracellular matrix, cells, collagen fibers and inorganic salts in the form of hydroxyapatite crystals [3]. Hence, an accurate scaffold designed to replace bone tissue must mimic the mineralized 3D structural composite to properly replicate its function. Several biocompatible materials have been developed to imitate the tubular 3D structure of bone including a biodegradable polymer fiber in a polymer matrix (polylactid acid fibers in a poly 1-caprolactone matrix) [4], hydroxyapatite ceramics [5] and polymer fibers filled with growth factors such as bone morphogenetic protein 7 [6]. Hydroxyapatite ceramics have been shown to permit osteoblast cell adhesion and proliferation [7], however, the use of pure ceramics as scaffold materials for tissue engineering is undesirable because of its brittleness and low strength [8]. The approach of creating polymer-ceramic composites can minimize the brittleness associated with pure ceramic. Thus the approach of creating such composites is optimum for creating biomimetic bone tissue. Our research group have developed and characterized the *in vitro* properties of a polymer-ceramic composite, using bacterial cellulose (BC) and physiological calcium-deficient hydroxyapatite (CdHA), an osteoconductive and bioresorbable mineral, for bone tissue regeneration [9, 10]. Furthermore, additional work from our group has shown that BC polymer scaffolds promote mesenchymal stem cell adhesion, proliferation and differentiation into osteocytes *in vitro* [11].

Native bone tissue is formed in layers of lamellae fiber bundles which contain near parallel arrays of collagen type I, a 1 nm diameter nanofibril structure, which forms 100-200 nm triple helical collagen molecules in the tissue matrix [3, 12]. In this paper, the near parallel arrays of collagen fiber in native bone are mimicked using BC scaffolds with oriented nanofibrils. BC scaffolds, a ~32 nm diameter nanofibril hydrogel, are very attractive materials for bone tissue engineering because they are chemically pure, biocompatible, moldable, cost-effective, possess high mechanical strength and large accessible surface area/volume ratio [9, 11, 13, 14]. BC is derived from a natural source and can be easily synthesized from non-pathogenic bacterium such as *Gluconacetobacter sp* into non-degradable, nanoporous scaffolds in the form of hydrogels [11]. Non-degradable BC have been used as a scaffold for treatment of second- or third-degree burns [15], chronic ulcers [16], and for artificial blood vessels [17]. Non-degradable BC can be rendered degradable through a periodate ring-opening oxidation, which degrades by simple hydrolysis mechanism [9, 14].

BC is a readily moldable scaffold which takes the shape of the vessel in which it is synthesized. We have previously made varying geometry of 3D BC by changing the shape of the dish used to synthesize the scaffold [11]. Recently, researchers have generated tubular BC (BC-TS) scaffolds with oriented fibers using oxygen permeable silicone tubes that are supplied with oxygen [18, 19]. BC-TS is beneficial for bone tissue engineering because the oriented fibers may be used to imitate the inherent oriented collagen fibers in native bone. Furthermore, in bone, CdHA minerals form along the oriented collagen fibers of the tissue [3]. As demonstrated in the previously studies, CdHA can be deposited in BC scaffolds to form BC-CdHA composites [9, 10]. Here, the structure and function of native bone tissue is mimicked, more specifically the collagen-CdHA composite in bone, by depositing osteoconductive and bioresorbable CdHA minerals in BC-TS with oriented fibers. Combined with osteocyte-forming mesenchymal stem cells, this biomimetic fibrous composite could be used for bone defect repair and reconstruction. We hypothesized that aligned mesenchymal stem cell morphology can be induced on the scaffolds due to the orientation of the fibers and as previously reported [20].

Successful bone repair approaches may include an osteoconductive scaffold that permits excellent cell adhesion and proliferation, and cells with an osteogenic potential. In this study, biocompatible and biodegradable BC-TS-CdHA composite scaffolds with various sizes were prepared and their morphological properties characterized. The ability of the scaffolds to support the proliferation, viability and osteocyte differentiation of equine-derived bone marrow mesenchymal stem cells (EqMSCs) *in vitro* was also evaluated for potential tissue engineering use.

MATERIALS AND METHODS

Materials

All chemicals, cell culture supplements and disposable tissue culture supplies were purchased from Sigma-Aldrich (Saint Louis, MO) unless otherwise noted. During the bone marrow extraction to obtain equine-derived bone marrow mesenchymal stem cell, animal procedures were carried out according to Institutional Animal Care and Use Committee protocol no. 1953.

Preparation of tubular bacterial cellulose and its composites

Silicone tubes (Fisher Scientific, Pittsburgh, PA), with inner diameters (ID) of 6.35 mm and 9.525 mm, wall thicknesses of 6.35 mm, and 100 mm in length were prepared for BC-TS synthesis. The tubes were washed, dried and sterilized in an autoclave (1 bar, 120 °C) for 30 min prior to use. Bacterial strain *Gluconacetobacter sucrofermentans* was commercially obtained from the American Type Culture Collection (Manassas, VA) (ATCC 700178). Pre-cultures of BC were made in a Schramm and Hestrin medium [21] and the constituents are described by Favi et al. [11]. For BC-TS production, pre-cultures of the bacteria were diluted 1:10 in fresh media and poured on the outside surface of the sterile silicone tubes (Fig. 1A) or on the inside surface the sterile tubes (Fig. 1B). This method of synthesizing BC-TS is a modification of a method previously described by Putra et al. [19]. The BC-TS were synthesized for 14 days and purified in distilled/deionized water using a method described by Favi et al. [11].

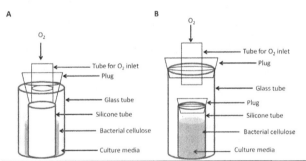

Figure 1: Schematic representation of BC-TS produced in silicone tubes. Illustration of BC-TS produced on the outside surface of the silicone tube (A) and on the inside surface of the silicone tube (B).

The oxidation of BC-TS (OBC-TS) was performed as previously described [9]. Briefly, BC-TS samples were placed in a capped vessel covered with aluminum foil that contained 50 mM $NaIO_4$ (sodium metaperiodate) in 5% *n*-propanol, and placed on an orbital shaker for 24 h at 23 °C. The oxidation reaction was stopped and excess periodate consumed by placing the vessel in an ice bath and adding 0.5 mL of glycerol. The OBC-TS were then purified with numerous changes of deionized water.

Calcium-deficient hydroxyapatite (CdHA) minerals were deposited within the unmodified BC-TS (BC-TS-CdHA) and oxidized BC-TS (OBC-TS-CdHA) by performing an alternating incubation cycle with calcium and phosphate solutions (modified from Hutchens et al. [10]). Briefly, BC-TS and OBC-TS are suspended in 5.0 mM $CaCl_2$ under agitation in an ortibal shaker for 24 h (23 °C), rinsing the samples briefly in deionized water, and then transferring the pellicle to 3.0 mM Na_2HPO_4 under agitation for another 24 h (23 °C) to obtain CdHA.

The morphology of the lyophilized BC-TS and its composites were examined using a scanning electron microscope (SEM). For lyophilization, hydrated samples were placed in a -80 °C freezer for 24 h and then lyophilized in a FreeZone® 4.5 Freeze Dry System (Labconco, Kansas City, MO) for at least 24 h at -50 °C and 1.0×10^{-3} Pa. The lyophilized samples were mounted on carbon tape and sputtered with gold on a Spi Module Sputter Coater (Spi Supplies: Westchester, PA, USA) at 20 mA for 30 s. The samples were then analyzed on a LEO 1525 SEM (Zeiss: Oberkochen, Germany).

Cell culture

Previously isolated equine-derived bone marrow mesenchymal stem cell (EqMSCs) were obtained by centrifugation of bone marrow aspirates from the sternum of a healthy 11-years-old male horse, characterized and cryo-preserved in our lab as previously described (Dhar et al.[22], Favi et. [11]). The cells were grown in Dulbecco's Modified Eagle Medium (DMEM) containing 10% fetal bovine serum (Hyclone, Logan, Utah) and 1% penicillin/streptomycin solution (P/S) (Invitrogen, Carlsbad, CA). Cell passages 2-6 were used for the experiments.

Cell seeding

BC-TS and its composites were prepared as circular discs for the cell study using 6 mm Miltex Inc. disposable biopsy punches (Fisher Scientific) and a 17.46 mm round hole arch punch (McMaster Carr, Atlanta, GA). For cell culture studies, the BC-TS and its composites were sterilized by autoclave (1 bar, 120 °C) for 30 min, and were pre-soaked in phenol red-free growth media for at least 24 h prior to cell seeding.

Cellular viability and proliferation assay: MTS test

Cellular viability and proliferation were assessed as previously described by Favi et al. [11]. Cell viability was assessed after 1, 2, and 3 days using the CellTiter 96® AQueous Non-Radioactive (MTS) assay (Promega, Madison, WI) according to the manufacturer's instructions. All experiments were simultaneously conducted in triplicate on BC-TS and its composites, which were placed in 96-well tissue culture plates (TCP). 2.5×10^4 cells were seeded on the scaffolds and incubated for 1, 2 and 3 days in the growth media at 37 °C/5% CO_2. EqMSCs seeded on the BC-TS and its composites were rinsed three times in phosphate buffer solution (PBS) and immersed in a mixture containing serum-free cell culture medium and MTS reagent in a 5:1 ratio and incubated for 3 h at 37 °C/5% CO_2. Then, 100 μL (n = 2) were transferred to 96-well plates and the optical density (O.D.) was measured on a microplate fluorescence reader (BioTek, Winooski, VT) using an absorbance of 490 nm. BC-TS scaffolds without cells and placed in the same media were used as blanks. The recorded absorbance values at 490 nm were subtracted from their respective blank readings to yield the corrected absorbance.

Cellular adhesion and cell viability staining using fluorescent microscopy

EqMSCs were seeded on BC-TS and its composites at a density of 2.5×10^4 cells/well. All experiments were conducted in 96-well TCP. Cell adhesion and viability was assessed after 1, 2 and 3 days using calcein-AM (Invitrogen, Eugene, OR) and propidium iodide (PI) (Invitrogen, Carlsbad, CA). Cells were stained according to the manufacturer's protocols and subsequently visualized using a Zeiss Axiovert 40C microscope (Carl Zeiss MicroImaging, Inc., Thornwood, NY) equipped with a Nikon Digital Sight DS-Qi1Mc camera (Nikon Instruments Inc., Melville, NY).

Cellular differentiation

Osteogenesis assay was performed as described by Favi et al. [11]. All experiments were conducted in 24-well TCP. Roughly 1.0×10^5 cells/well were seeded in media on BC-TS and its composites. 70–80% confluent cells were induced to differentiate into osteocytes. Osteogenic differentiation was induced in growth media supplemented with 100 nM dexamethasone, 10 mM β-glycerophosphate, and 0.25 mM ascorbic acid. Differentiation cell groups (induced cells with osteogenic media) were monitored with control cell groups (non-induced cells without osteogenic media). Media was changed every 2-3 days, and differentiation was assessed using alizarin red (Fisher Scientific) staining after 8 days. Cells were fixed in 4% paraformaldehyde (PFA) for 10 min at room temperature and stained with alizarin red for 30 min for the detection of calcium in differentiated cells. Images were taken with a Zeiss Axiovert 40C microscope equipped with a Nikon Digital Sight DS-Fi2 camera.

Statistical analysis

Data are expressed as mean ± standard deviation (SD) of at least three independent samples. Statistical comparisons between groups were performed with a two-tailed Student's t-test, $p < 0.05$ was considered significant.

RESULTS AND DISCUSSION

BC-TS samples were successfully generated in oxygen-permeable silicone tubes. Four different sizes of scaffolds were generated from the synthesis of *Gluconacetobacter sucrofermentans* in a Schramm and Hestrin medium. BC-TS scaffolds synthesized on the inner surface of the 6.35 mm ID silicone tubes (8.3 mm diameter, 1.6 mm thickness) produced scaffolds whose diameter were smaller than those produced on the outer surface of the equivalent size tube (20 mm diameter, 1.0 mm thickness) (Fig. 2C). Similar increase in scaffold diameter were also observed with the BC-TS scaffolds synthesized on the inner surface of the 9.525 mm ID silicone tubes (13.5 mm diameter, 1.1 mm thickness) (Fig. 2B) compared to the outer surface (24.2 mm diameter, 1.7 mm thickness) of the equivalent size tube (Fig. 2D).

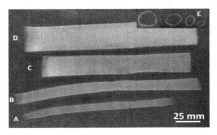

Figure 2: Photographs of BC-TS scaffolds synthesized on the inside surface (A) and on the outside surface (C) of 6.35 mm ID silicone tubes. Photographs of BC-TS scaffolds synthesized on the inside surface (B) and on the outside surface (D) of the 9.525 mm ID silicone tubes. Top view image (E) of the four sizes of BC-TS hydrogels synthesized using silicone tubes.

SEM images of the 8.3 mm BC-TS produced under various treatments demonstrated that aligned cellulose fibers were successfully generated in the scaffolds (Fig. 3). SEM images of the BC-TS before (Fig. 3A) and after (Fig. 3B) periodate oxidation showed that the scaffolds maintained their morphological integrity of the nanofibers during the chemical reaction. CdHA ceramics were successfully deposited in the non-oxidized scaffold (Fig. 3C) and oxidized scaffold (Fig. 3D) as illustrated in the SEM images.

MTS assay analysis was performed to determine the viability of EqMSCs on the tubular BC and its composites. MTS assay analysis results of EqMSCs seeded on the tubular BC and its composites are shown in Fig. 4. EqMSCs seeded on OBC-TS and OBC-TS-CdHA showed a significant increase compared to BC-TS and BC-TS-CdHA during the three days in culture. Furthermore, the stem cells demonstrated a metabolic rate of increase during the three days in culture signifying that the cells were viable and proliferating on the scaffolds.

Calcein-AM and PI staining was performed to visualize the viability of the cells on the scaffolds. After 1 day in culture (Fig. 5), the cells on all the tubular BC scaffolds were viable. However, the cells on the oxidized scaffolds (Fig. 5B, 5D) showed distinct mesenchymal stem cell phenotype of full spread-out morphology on the scaffolds. The cells on the non-oxidized scaffolds (Fig. 5A, 5C) showed less elongated shapes and their morphology was round. We had previously illustrated that up to day 7 in culture, EqMSCs seeded on non-oxidized BC are round in shape [11]. Additionally, EqMSCs do proliferate over time and by day 7 and day 14 demonstrate the full spread-out morphology on non-oxidized BCs [11].

EqMSCs successfully differentiated into osteocytes on the tubular BC scaffolds *in vitro* (Fig. 6). The cells exhibited the potential to differentiate into bone cells on the induction media as indicated by the detection of calcium in the differentiated cells (Fig. 6E-6H). These results show that BC-TS and its composites enable EqMSCs adhesion, proliferation and differentiation into osteocytes.

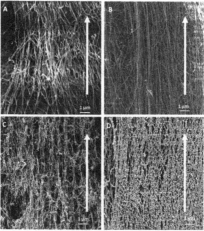

Figure 3: SEM images of 8.3 mm diameter BC-TS prepared using various treatments. SEM images of BC-TS (A), OBC-TS (B), BC-TS-CdHA (C) and OBC-TS-CdHA. Arrows indicate the direction of the longitudinal axis of the silicone tube during scaffold synthesis.

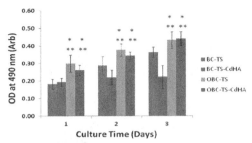

Figure 4: Cellular viability assay: MTS test. Comparison of proliferation of cells as determined by MTS assay for EqMSCs seeded on BC-TS, OBC-TS, BC-TS-CdHA and OBC-TS-CdHA scaffolds for 1, 2 and 3 days. OBC-TC and OBC-TS-CdHA showed a significant increase compared to BC-TS and BC-TS-CdHA during the three days in culture (p<0.05).

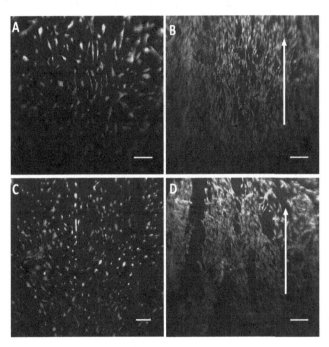

Figure 5: Cellular adhesion and cell viability stained with calcein-AM and PI using fluorescent microscopy. Cell viability of EqMSCs seeded on BC-TS (A), OBC-TS (B), BC-TS-CdHA (C) and OBC-TS-CdHA (D) after 1 day in culture. Cells were analyzed by calcein-AM which exhibits green fluorescence and demonstrates live cells and PI which displays red fluorescence and demonstrates dead cells. Fluorescent micrographs showed that the cells were viable on the scaffolds. Arrows in image show the aligned direction in which the cells grew following the orientation of the cellulose fibers of tubular BC. Scale bar = 100 μm.

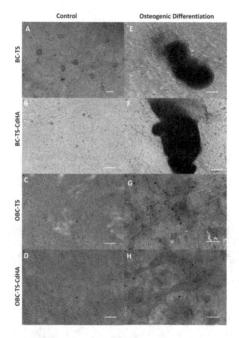

Figure 6: Osteogenesis differentiation capacity of EqMSCs on the tubular BC and its composites after 8 days of in vitro differentiation. Osteogenesis was induced using the β-glycerophosphate-based method and was demonstrated by the detection of calcium in the mineralized matrix indicated by alizarin red stain shown in E, F, G and H (A, B, C and D: non-induced controls). Scale bars = 100 μm.

CONCLUSIONS

BC-TS is a natural hydrogel scaffold that was successfully prepared, and the scaffold and its composites were characterized using SEM. The lyophilized BC-TS and its composites illustrated aligned nanofibrous morphology. It was demonstrated that the BC-TS scaffold and its composites were cytocompatible with EqMSCs *in vitro*. BC-TS scaffolds and its composites supported the adhesion, proliferation and osteogenic differentiation of EqMSCs. BC-TS scaffolds and its composites are promising alternative scaffolds for bone tissue engineering.

ACKNOWLEDGMENTS

This study was supported by the College of Veterinary Medicine's Center of Excellence grant and the National Institute of Health Training Grant (NIH/ NIGMS-IMSD: R25 GM086761).

REFERENCES

[1] Lohmander LS. Tissue engineering of cartillage: Do we need it, can we do it, is it good and can we prove it? In: Bock G, Goode J, editors. Tissue Engineering of Cartilage and Bone: Novartis Foundation Symposium 249. Chichester: John Wiley & Sons Ltd; 2003. p. 2-10.

[2] Marolt D, Knezevic M, Novakovic GV. Bone tissue engineering with human stem cells. Stem Cell Res Ther. 2010;1:10-20.

[3] Cui D, Daley W, Naftel JP, Lynch JC, Haines DE, Yang G, et al. Atlas of Histology: With Functional and Clinical Correlations. 1st ed. Philadelphia: Wolters Kluwer Health/Lippincott Williams & Wilkins; 2011.

[4] Guarino V, Urciuolo F, Alvarez-Perez MA, Mele B, Netti PA, Ambrosio L. Osteogenic differentiation and mineralization in fibre-reinforced tubular scaffolds: theoretical study and experimental evidences. J R Soc Interface. 2012;9:2201-12.

[5] Ustundag CB, Kaya F, Kamitakahara M, Kaya C, Ioku K. Production of tubular porous hydroxyapatite using electrophoretic deposition. J Ceram Soc Jpn. 2012;120:569-73.

[6] Berner A, Boerckel JD, Saifzadeh S, Steck R, Ren J, Vaquette C, et al. Biomimetic tubular nanofiber mesh and platelet rich plasma-mediated delivery of BMP-7 for large bone defect regeneration. Cell Tissue Res. 2012;347:603-12.

[7] Akkouch A, Zhang Z, Rouabhia M. Engineering bone tissue using human dental pulp stem cells and an osteogenic collagen-hydroxyapatite-poly($\square$-lactide-co-ε-caprolactone) scaffold. J Biomater Appl. 2013.

[8] Kim B-S, Kang HJ, Lee J. Improvement of the compressive strength of a cuttlefish bone-derived porous hydroxyapatite scaffold via polycaprolactone coating. J Biomed Mater Res B. 2013;101:1302-9.

[9] Hutchens S, Benson R, Evans B, Rawn C, O'Neill H. A resorbable calcium-deficient hydroxyapatite hydrogel composite for osseous regeneration. Cellulose. 2009;16:887-98.

[10] Hutchens SA, Benson RS, Evans BR, O'Neill HM, Rawn CJ. Biomimetic synthesis of calcium-deficient hydroxyapatite in a natural hydrogel. Biomaterials. 2006;27:4661-70.

[11] Favi PM, Benson RS, Neilsen NR, Hammonds RL, Bates CC, Stephens CP, et al. Cell proliferation, viability, and in vitro differentiation of equine mesenchymal stem cells seeded on bacterial cellulose hydrogel scaffolds. Mater Sci Eng C Mater Biol Appl. 2013;33:1935-44.

[12] Bigg HF, Rowan AD, Barker MD, Cawston TE. Activity of matrix metalloproteinase-9 against native collagen types I and III. FEBS J. 2007;274:1246-55.

[13] Helenius G, Bäckdahl H, Bodin A, Nannmark U, Gatenholm P, Risberg B. In vivo biocompatibility of bacterial cellulose. J Biomed Mater Res A. 2006;76A:431-8.

[14] Bielecki S, Krystoynowicz A, Turkiewicz M, Kalinowska H. Bacterial cellulose. In: Vandamme EJ, De Baets S, Steinb A, editors. Biopolymers: Vol 5, Polysaccharides I, Polysaccharides from Prokaryotes. Weinham: Wiley; 2001. p. 37–46.

[15] Fontana J, De Souza A, Fontana C, Torriani I, Moreschi J, Gallotti B, et al. Acetobacter cellulose pellicle as a temporary skin substitute. Applied Biochemistry and Biotechnology. 1990;24-25:253-64.

[16] Kucharzewski M, Slezak A, Franek A. Topical treatment of non-healing venous leg ulcers by cellulose membrane. Phlebologie. 2003;32:138-69.

[17] Klemm D, Schumann D, Udhardt U, Marsch S. Bacterial synthesized cellulose — artificial blood vessels for microsurgery. Prog Polym Sci. 2001;26:1561-603.

[18] Wang J, Wan Y, Huang Y. Immobilisation of heparin on bacterial cellulose-chitosan nano-fibres surfaces via the cross-linking technique. IET Nanobiotechnol 2012 6:52-7.

[19] Putra A, Kakugo A, Furukawa H, Gong JP, Osada Y. Tubular bacterial cellulose gel with oriented fibrils on the curved surface. Polymer. 2008;49:1885-91.

[20] Lyu S, Huang C, Yang H, Zhang X. Electrospun fibers as a scaffolding platform for bone tissue repair. J Orthop Res. 2013;31:1382-9.

[21] Schramm M, Hestrin S. Factors affecting Production of Cellulose at the Air/ Liquid Interface of a Culture of Acetobacter xylinum. J Gen Microbiol. 1954;11:123-9.

[22] Dhar M, Neilsen N, Beatty K, Eaker S, Adair H, Geiser D. Equine peripheral blood-derived mesenchymal stem cells: Isolation, identification, trilineage differentiation and effect of hyperbaric oxygen treatment. Equine Vet J. 2012;44:600-5.

Mater. Res. Soc. Symp. Proc. Vol. 1621 © 2014 Materials Research Society
DOI: 10.1557/opl.2014.70

Dielectrophoretical fabrication of hybrid carbon nanotubes-hydrogel biomaterial for muscle tissue engineering applications

Javier Ramón-Azcón[1], Samad Ahadian[1], Raquel Obregon[2], Hitoshi Shiku[2], Ali Khademhosseini[1,4], Tomokazu Matsue[1,2]

[1]WPI-Advanced Institute for Materials Research, Tohoku University, Sendai 980-8577, Japan
[2]Graduate School of Environmental Studies, Tohoku University, Sendai 980-8579, Japan
[3]Department of Medicine, Center for Biomedical Engineering, Brigham and Women's Hospital, Harvard Medical School, Cambridge, Massachusetts 02139, USA; Harvard–MIT Division of Health Sciences and Technology, Massachusetts Institute of Technology, Cambridge, Massachusetts 02139, USA; Wyss Institute for Biologically Inspired Engineering, Harvard University, Boston, Massachusetts 02115, USA; Department of Maxillofacial Biomedical Engineering and Institute of Oral Biology, School of Dentistry, Kyung Hee University, Seoul 130-701, Republic of Korea.

ABSTRACT

Dielectrophoresis (DEP) approach was employed to achieve highly aligned multi-walled carbon nanotubes (MWCNTs) within the gelatin methacrylate (GelMA) hydrogels in a facile, rapid, inexpensive, and reproducible manner. This approach enabled us to make different CNTs alignments (e.g., vertical or horizontal alignments) within the GelMA hydrogel using different electrode designs or configurations. Anisotropically aligned GelMA-CNTs hydrogels showed considerably higher conductivity compared to randomly distributed CNTs dispersed in the GelMA hydrogel and the pristine and non-conductive GelMA hydrogel. Adding 0.3 mg/mL CNTs to the GelMA hydrogel led to a slight increase in the mechanical properties of the GelMA and made it to behave as a viscoelastic material. Therefore, it can be used as a suitable scaffold for soft tissues, such as skeletal muscle tissue. 3D microarrays of skeletal muscle myofibers were then fabricated based on the GelMA and GelMA-CNTs hydrogels and they were characterized in terms of gene expressions related to the muscle cell differentiation and contraction. Owing to high electrical conductivity of aligned GelMA-CNTs hydrogels, the engineered muscle tissues cultivated on these materials demonstrated superior maturation and functionality particularly after applying the electrical stimulation (voltage 8 V, frequency 1 Hz, and duration 10 ms for 2 days) compared to the corresponding tissues obtained on the pristine GelMA and randomly distributed CNTs within the GelMA hydrogel.

INTRODUCTION

In vitro generation of tissues in principle involves the proliferation and differentiation of cells within a scaffold mimicking the formation of native tissues within the extracellular matrix (ECM). Hydrogels have been considered as ideal materials to mimic the ECM due to high water content and biodegradability [1,2]. However, they generally have weak mechanical properties and low conductivity, which have limited their ability to engineer soft and electroactive tissues, such as skeletal muscle, cardiac, and neural tissues. Therefore, it is desirable to control and tune the mechanical and electrical properties of hydrogels for soft tissue engineering applications. Nanomaterials have recently gained much attention to address these problems [3]. For instance, hydrogels impregnated with gold nanostructures were demonstrated to improve the electrical conductivity and cellular excitability of both cardiomyocyte and neural cells [4,5]. Even though, gold nanowires enhanced the electrical conductivity of alginate scaffolds, it was reported that they could not improve the elastic modulus of scaffolds [4]. In addition, as reported by the authors, there was no thoroughly interconnected structure of gold nanowires within the scaffolds. As a result, the long-range propagation of current was not occurred. Poly(lactic-*co*-glycolic-acid) (PLGA) hydrogels incorporating carbon nanofibers were demonstrated to have higher conductivity compared to the pure PLGA hydrogels [6]. The composite PLGA-carbon nanofiber hydrogels also enhanced the adhesion and growth of both cardiomyocyte and neural cells. It was recently demonstrated that the gelatin methacrylate (GelMA) hydrogel was able to emulate the ECM for various cell types such as muscle, cardiac, and endothelial cells [7]. However, increasing the mechanical properties of GelMA hydrogels through concentration and molecular weight limit their cell viability and growth, morphogenesis, and cell migration [8].

CNTs show most of their unique properties in the tube axis direction. An approximation showed that unidirectionally aligned CNTs in a material could substantially improve electrical and mechanical properties of composite material along the CNTs direction even at low concentrations (*i.e.*, 0.1 wt%) [9]. In particular, the anisotropic conductivity can be achieved as the CNTs are aligned [10]. Therefore, it is desired to establish methods to fabricate well-aligned CNTs within the composite materials [11]. Here, dielectrophoresis (DEP) method was used to align the CNTs within the GelMA hydrogel. DEP stands as a powerful technique to manipulate the particles based upon their interactions with an AC electric field by which a charge polarization is occurred within the particles and their surrounding medium [12,13].

DISCUSSION

The aim of this work was to improve mechanical and electrical properties of GelMA hydrogel as a conventional scaffold for fabricating the engineered tissues. Internal nanostructured network of CNTs was made within the GelMA hydrogel architected by the DEP forces leading to improvements in both electrical and mechanical properties of pristine GelMA hydrogel. The performance of this novel scaffold was evaluated to engineer the skeletal muscle tissues.

Fabrication of nanostructured GelMA-CNT hydrogels

The DEP approach was used to align CNTs within the GelMA hydrogel (Figure 1). GelMA was synthesized as previously described [7]. DEP is a useful technique by which the dielectric particles in a medium are polarized due to applying an AC electric field. Indeed, the AC electric field induced the dipole moment within the CNTs and pushed them to be aligned along with the electric field direction. In particular, in this work, it was used semi-metallic MWCNTs with carboxylic groups on their surface to facilitate their dispersion in hydrophilic solutions. In addition, pristine MWCNTs have increased the conductivity of biomaterials because of their semi metallic properties [14].

It has been experimentally demonstrated that the MWCNTs can be dielectrophoretically manipulated as the frequency increases from 2 kHz to 2 MHz. The main objective in this work was to dielectrophoretically manipulate the MWCNTs and consequently align them within the GelMA hydrogel. Therefore, AC electric field was applied at high frequencies. At high frequencies and voltage (2 MHz and 20 Vpp), it was successfully patterned the MWCNTs within the GelMA hydrogel using IDA-Pt electrodes. This nanostructure was specifically designed to cultivate and electrically stimulate skeletal muscle tissue as will be described later. Here, it is reported two additional composite nanostructures with vertically aligned CNTs within the GelMA hydrogel. These structures were easily obtained upon applying the AC electric field within the bottom and top electrodes (frequency 2 MHz and voltage 20 Vpp). Similar nanostructures have been reported in the literature for the biomedical applications, such as vertically aligned carbon nanofibers (CNFs) [15]. The CNFs consisted of multi-walled graphene structures stacked on top of each other, like a stack of ice cream cones [16]. These vertically aligned structures have been widely used as neuronal interfaces, gene delivery arrays, and bio-sensors [16]. Generally, the CNFs were synthesized *via* the plasma enhanced chemical vapour deposition technique and were aligned perpendicular to a conductive substrate. This process is long and requires several steps with high vacuum process and gas chemical reactions. Instead, DEP approach is able to make vertically CNTs nanostructure in few minutes within the GelMA hydrogel. This is an exciting research area to make different patterns of vertically aligned CNTs in a simple and rapid manner to use in the biological interfaces and related bio-devices.

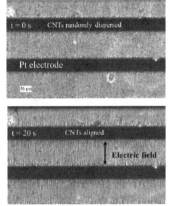

Figure 1. Process of the carbon nanotubes (CNTs) alignment as the bundle within hydrogel under dielectrophoresis (DEP) force.

Conductivity features of the GelMA-CNTs hydrogels

GelMA-CNTs composites with the CNTs loading of 0.1, 0.3, 1, and 2 mg/mL were prepared with and without CNTs alignment. DC conductivity of all these structures were measured and compared with pristine GelMA. As it can be seen in Figure 2, conductivity was dramatically increased as a function of CNTs concentration and importantly there was a significant increase in the conductivity while the CNTs were aligned within the GelMA hydrogel compared to the random CNTs within the GelMA. Similar trend was observed for the impedance measurements within the hydrogels (Figure 2-C). Impedance spectra were acquired for the GelMA and GelMA-CNTs hydrogels on the IDA-Pt electrodes over a frequency range from 10 to 10^5 Hz with the perturbation amplitude of 25 mV. Higher conductivity of GelMA hydrogels with aligned CNTs compared to those involving random CNTs may be due to the formation of continuous current passages along with the electric field direction. It seems that the hydrogel wrapping around the CNTs minimizes the electron transfer efficiency between the nanotubes; however, it has little effect on the dielectrophoretically aligned CNTs.

Mechanical features of GelMA-CNTs hydrogels

AFM is a potent tool in order to image the topography of solid surfaces at the nanoscale. It also offers the potential for measuring the mechanical properties of materials with nano- or microscale resolution. AFM was successfully applied for the characterization of MWCNTs/elastomer composites while the CNTs were homogenously distributed in the polymer at different concentration. In this work, a micromechanical mapping technique was used to measure the mechanical properties of the nanostructured GelMA-CNTs composites, including elasticity, adhesive energy, and surface topography. The results as summarized in Figure 3 show that the Young`s modulus for the pristine GelMA was 12.5 kPa in agreement with our previously reported value derived by the conventional stress-strain measurements [7]. As the GelMA was reinforced by the CNTs, its Young`s modulus increased to 20.9 and 23.4 kPa for the aligned and random CNTs within the GelMA hydrogel, respectively. This increase in the Young`s modulus was likely due to the formation of a well-connected three-dimensional network structure inside the hydrogel.

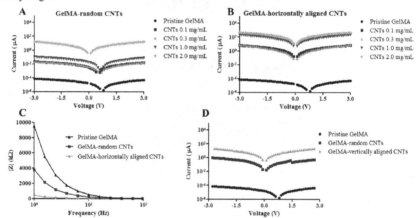

Figure 2. Electrical features of GelMA-CNT composites with different concentrations of CNTs (0.1, 0.3, 1.0, and 2.0 mg/mL), along with the pristine GelMA (*i.e.*, 0 mg/mL CNTs). (A,B and D) I-V curves. (C) Impedance measurements of the pristine GelMA and the composites loaded with 0.3 mg/mL CNTs. The perturbation amplitude was 25 mV.

Skeletal muscle tissue generation using the GelMA-CNTs hydrogels

Muscle cells were aligned along with the CNTs direction inside the GelMA hydrogels. This structure also allows to maximize the interactions between C2C12 muscle cells and MWCNTs compared to the 2D structure. Therefore, it plays a modulator role to promote the C2C12 cells differentiation and tissue maturation. After myotubes differentiation, the skeletal muscle tissues were stimulated through the IDA-Pt electrode as was described in our previous work. Here, impregnation of GelMA hydrogels with CNTs and particularly with aligned CNTs led to the promotion in the differentiation of muscle cells probably as a result of: a) high electrical conductivity of CNTs. It seems that tight contact and interaction between CNTs and muscle cell membrane may change the electric potential of cell membranes and decrease the electrical impedance as observed for CNTs in close contact with cardiac and neural cells; b) providing suitable anchoring sites for the cells to enhance the adhesion and differentiation of muscle cells. Note that the anchorage-dependent cells such as muscle cells need good adhesion to the substrate to attach, grow, and maintain the cellular function.

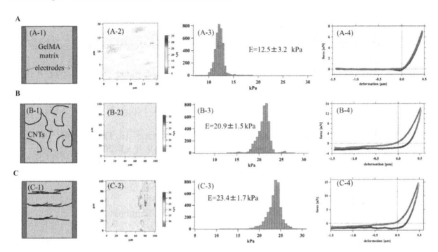

Figure 3. Mechanical and morphological properties of the 5% GelMA and aligned/non-aligned GelMA-CNTs composites (0.3 mg/mL CNTs), as measured by AFM. (A) Schematic representation, Young's modulus map and its histogram, and force deformation curves for the pure 5% GelMA hydrogel (A-1, A-2, A-3, and A-4, respectively). The red curve represents the force deformation curve as the AFM cantilever approached the surface, while the blue curve is the force deformation curve produced when the cantilever left the surface. The red line was calculated according to the DMT theory. (B) Schematic representation, Young's modulus map and its histogram, and force deformation curves of the CNTs randomly dispersed in the GelMA hydrogel (B-1, B-2, B-3, and B-4, respectively). (C) Schematic representation, Young's modulus map and its histogram, and force deformation curves of the CNTs horizontally aligned within the GelMA hydrogel (C-1, C-2, C-3, and C-4, respectively).

CONCLUSIONS

DEP approach was proposed to achieve highly aligned MWCNTs within the GelMA hydrogels in a facile and rapid way. Anisotropically aligned GelMA-CNTs hydrogels showed considerably higher conductivity compared to randomly distributed CNTs dispersed in the GelMA hydrogel and pristine and non-conductive GelMA hydrogel. The latter hydrogels showed a viscoelastic behavior that is suitable for the soft tissue engineering applications. 3D microarrays of skeletal muscle myofibers were then fabricated based on the GelMA and GelMA-CNTs hydrogels. Owing to high electrical conductivity of aligned GelMA-CNTs hydrogels, engineered muscle tissues cultivated on these materials demonstrated superior maturation and functionality particularly after applying the electrical stimulation compared to the corresponding tissues obtained on the pristine GelMA and randomly distributed CNTs within the GelMA hydrogel.

ACKNOWLEDGMENTS

This work was supported by the World Premier International Research Center Initiative (WPI), MEXT, Japan.

REFERENCES

1.   Slaughter B V, Khurshid SS, Fisher OZ, Khademhosseini A, Peppas NA., Adv. Mater. 21, 3307 (2009).
2.   Peppas NA, Hilt JZ, Khademhosseini A, Langer R., Adv. Mater. 18, 1345 (2006).
3.   Dvir T, Timko BP, Kohane DS, Langer R. Nat., Nanotechnol. 6, 13 (2011).
4.   Dvir T, Timko BP, Brigham MD, Naik SR, Karajanagi SS, Levy O, et al., Nat. Nanotechnol. 6, 720 (2011).
5.   You J-O, Rafat M, Ye GJC, Auguste DT., Nano Lett. 11, 3643 (2011).
6.   Stout DA, Basu B, Webster TJ., Acta Biomater. 7, 3101 (2011).
7.   Nichol JW, Koshy ST, Bae H, Hwang CM, Yamanlar S, Khademhosseini A, Biomaterials 31, 5536 (2010).
8.   Aubin H, Nichol JW, Hutson CB, Bae H, Sieminski AL, Cropek DM, et al., Biomaterials 31, 6941 (2010).
9.   MacDonald RA, Voge CM, Kariolis M, Stegemann JP, Acta Biomater. 4, 1583 (2008).
10.   Bal S., Materials and Design 31, 2406 (2010).
11.   Wang B, Ma Y, Li N, Wu Y, Li F, Chen Y, Adv. Mater. 22, 3067 (2010).
12.   Ramón-Azcón J, Yasukawa T, Lee HJ, Matsue T, Sánchez-Baeza F, Marco M-P, et al., Biosens. Bioelectron. 25, 1928 (2010).
13.   Ramón-Azcón J, Kunikata R, Sanchez F-J, Marco M-P, Shiku H, Yasukawa T, et al., Biosens. Bioelectron. 24, 1592 (2009).
14.   Voge CM, Johns J, Raghavan M, Morris MD, Stegemann JP, J. Biomed. Mater. Res. A 101, 231 (2013).
15.   Anatoli V Melechko, Ramya Desikan, Timothy E McKnight, Kate L Klein, Philip D Rack., J. Phys. D App. Phys. 42, 193001 (2009).
16.   Nguyen-Vu TDB, Chen H, Cassell AM, Andrews R, Meyyappan M, Li J., Small. 2, 89 (2006).

Mater. Res. Soc. Symp. Proc. Vol. 1621 © 2014 Materials Research Society
DOI: 10.1557/opl.2014.68

In Vitro Examination of Poly(glycerol sebacate) Degradation Kinetics: Effects of Porosity and Cure Temperature

Nadia M. Krook,[1] Courtney LeBlon,[2] and Sabrina S. Jedlicka[1,3,4]
Lehigh University[1] Materials Science & Engineering,[2] Mechanical Engineering & Mechanics,[3]
Bioengineering Program,[4] Center for Advanced Materials & Nanotechnology,
Bethlehem, PA, 18015, U.S.A

ABSTRACT

Poly(glycerol sebacate) (PGS) is a biodegradable and biocompatible elastomer that has been used in a wide range of biomedical applications. While a porous format is common for tissue engineering scaffolds, to allow cell ingrowth, PGS degradation has been primarily studied in a nonporous format. The purpose of this research was to investigate the degradation of porous PGS at three frequently used cure temperatures: 120°C, 140°C, and 165°C. The thermal, chemical, mechanical, and morphological changes were examined using thermogravimetric analysis, differential scanning calorimetry, Fourier transform infrared spectroscopy, compression testing, and scanning electron microscopy. Over the course of the 16-week degradation study, the samples' pores collapsed. The specimens cured at 120°C demonstrated the most degradation and became gel-like after 16 weeks. Thermal changes were most evident in the 120°C and 140°C cure PGS specimens, as shifts in the melting and recrystallization temperatures occurred. Porous samples cured at all three temperatures displayed a decrease in compressive modulus after 16 weeks. This *in vitro* study helped to elucidate the effects of porosity and cure temperature on the biodegradation of PGS and will be valuable for the design of future PGS scaffolds.

INTRODUCTION

Biocompatible and biodegradable poly(glycerol sebacate) (PGS) exhibits similar mechanical properties to those of soft body tissues [1,2]. Porous and nonporous PGS has been used in various applications of bioengineering, including drug delivery [3] and tissue engineering scaffolds to allow cell ingrowth [4-9]. Previous *in vivo* biodegradation of nonporous PGS shows that there was complete absorption in 60 days [1]. *In vitro*, nonporous PGS is almost completely degraded in 4 months in simulated body fluid (SBF) [10]. Very little is known about the degradation of porous PGS and data is limited to weight loss, chemical changes, and morphology for nonporous PGS [1,2,10]. Further *in vitro* research on the biodegradation will provide insight to the effects of degradation on material properties for better use of PGS in its biological applications.

The purpose of this research was to investigate the degradation effects of SBF on porous PGS at three different cure temperatures: 120°C, 140°C, and 165°C. There is also a direct link to changing cure temperatures and differences in mechanical properties. Increased cure temperature of PGS correlates with an increased amount of cross-linking and thus a greater elastic modulus [1]. This research is designed to better understand the effects of SBF on the properties of biocompatible and biodegradable porous PGS and the specific effect of the pores.

EXPERIMENT

Prepolymer Synthesis

A prepolymer was made of equimolar (1:1) amounts of anhydrous glycerol and sebacic acid (purified via recrystallization in ethanol 3 times) [1]. The particles were mixed in an airtight glass jar and gradually heated up to 120 °C under nitrogen gas flow while being stirred with a rotor (50 rpm) at 120 °C for 24 hours. The gas flow was then stopped and vacuum applied at –20 kPa for a further 24 hours to complete the process.

Specimen Preparation

Prepolymer was dissolved in THF to make 10% of a mixture with 90% NaCl particles. The pore size was determined by the NaCl particles sizes, which were sifted to be greater than greater than 177 μm and then added to the THF solution with prepolymer. Samples were cured at 120°C, 140°C, and 165°C for a week long each. Once the porous samples were cured, they were soaked in double distilled H_2O (ddH_2O) three times for 24 hours each to leach out the salt. Specimens were trimmed to a nearly constant weight and size, within ±10% of 0.1829 g and 12.7x38.1x3 mm, respectively. The samples were then sterilized under UV light for 48 hours.

***In Vitro* Degradation**

The biodegradation of the porous PGS was carried out in SBF [11]. SBF (pH 7.4) was made by dissolving the following reagents in 1L ddH_2O:137.51μM NaCl, 4.19μM NaHCO$_3$, 3.02μM KCl, 1.37μM K$_2$HPO$_4$ 3H$_2$0, 1.53μM MgCl$_2$•6H$_2$0, 20ml 1.0M HCl, 2.64μM CaCl$_2$, 506.90 nM Na$_2$SO$_4$, and 50.56μM TRIS Base.[4] Approximately 12 mL of SBF was placed in the test tubes with each sample. The samples were stored in an incubator at 37°C for the duration of the biodegradation experiment (16 weeks).

Every 2 weeks, 3 samples for each cure temperature were removed from the incubator and rinsed three times with ddH_2O. They were then left to dry under a sterile tissue culture hood for 24 hours and kept sterile until testing. The specimens were refrigerated before and after performing assays. The specimens were digitally imaged and analyzed using Thermogravimetric Analysis (TGA), Differential Scanning Calorimetery (DSC), Fourier Transform-Infrared (FTIR) Spectroscopy, and Compression Testing (ARES Rheometer). Scanning electron microscopy (SEM) was performed on select samples.

Methods

The chemical, mechanical, and thermal properties, as well as morphological changes were examined at 2-week intervals of the 16-week biodegradation study. A TA Instruments Q500 Thermogravimetric Analyzer was used for thermogravimetric analysis. Each sampled weighed approximately 12 mg. The analysis ran from a temperature of 30.00°C to 800.00°C with a ramp rate of 10.00°C/min and a nitrogen flow rate of 40mL/min. Three TGA data points are taken per cure temperature every two weeks.

DSC was performed using a DSC 2920 Differential Scanning Calorimeter. The weight of the samples ranged from 3-5 mg and secured in hermetic pans. The experiments equilibrated at

-60.00°C and ramped to 100.00°C at a rate of 10.00°C/min. One DSC run was taken per cure temperature at every timepoint.

Attenuated Total Reflectance-Fourier Transform Infrared Spectroscopy (ATR-FTIR) was completed on a Perkin Elmer Spectrum 100. The absorbance peaks were obtained during FTIR analysis where 16 scans were compiled for each sample. The scans ranged from 4000-650 cm⁻¹ with a 4 cm⁻¹ resolution. Three scans were taken per cure temperature at every timepoint.

Compression testing on an ARES rheometric system was executed with parallel plates at a rate of 0.083 mm/s for 30 seconds. The specimens were sectioned to a 3 mm size cube. Fifteen compression tests were averaged per cure temperature for every two weeks.

DISCUSSION

Morphological Changes

All samples began at a similar shape and size to obtain a mean weight within ±10% of each other. The most significant change occurred in the 120°C cured porous samples. Nearing 16 weeks, the integrity of specimens diminished until only small pieces remained that nearly matched the viscous prepolymer state, indicating significant degradation. The 140°C cured porous samples maintained their shape with a slight shrinkage in size by16 weeks, demonstrating moderate macroscopic degradation. Macroscopically there was no apparent change to the 160°C cured porous PGS specimens.

Scanning electron microscopy (SEM) techniques were implemented to observe microscopic changes in illustrative samples at week 2 and week 12 for 120°C and 160°C cure temperatures. The 120°C series of microscopic images can be seen in Figure 1a and 1b the 165°C series in Figure 1c and 1d. The SEM images show pore shrinkage by week 12 especially in 120°C cure. These images further demonstrate surface erosion, the degradation mechanism of PGS.

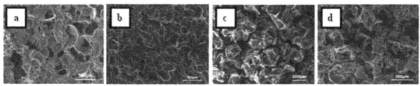

Figure 1. Scanning electron microscopy images of (a) 120°C cure specimens at week 2, (b) 120°C cure specimens at week 12, (c) 165°C cure specimens at week 2, and (d) 165°C cure specimens at week 12 time points.

ATR-FTIR

ATR-FTIR monitored chemical changes of the porous PGS samples during biodegradation. The peaks of interest in the FTIR spectrums are the carboxylic acid bend and hydroxyl stretch. In the carboxylic acid bend, the porous PGS specimens show an absorbance peak for all three cure temperatures at approximately 1730 cm⁻¹, representing the C=O double bond seen in Figure 2a. The compilation of the FTIR scans in Figure 2a for the prepolymer and all three of the cure temperatures from week 0 show that increasing cure temperature corresponds to an increased percent transmittance. In addition to the 1730 cm⁻¹ peak, PGS prepolymer reveals a peak at about

89

1700 cm^{-1} for the unsaturated carboxylic acid. This is shown in Figure 2b with the 120°C porous summary for the C=O peak. The porous 120°C cure PGS samples begin to develop the 1700 cm^{-1} peak by week 16 shown in Figure 2b. This transition is evidence of the breakdown of PGS crosslinks. Little change is observed in the PGS 140°C or 165°C cure temperature conditions.

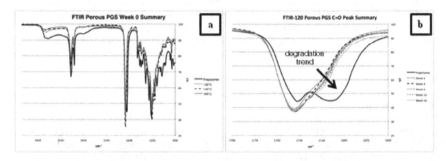

Figure 2. FTIR data shows (a) the summary of 120°C, 140°C, and 165°C week 0 with the prepolymer spectra and the featured FTIR spectra of the carboxylic acid bend for the 120°C cure porous PGS samples as a function of degradation time including the prepolymer scan.

DSC

The presented curves for the DSC data in Figure 3 are from the week 0 time points of the 120°C, 140°C, and 160°C cured porous PGS. The curves indicate that before degradation the effect of increasing cure temperatures is shorter peaks that are shifted to the left. A summary of the changes in crystallinity (T_C), melting (T_M), and transition glass temperatures (T_g), can be seen in Table I. During degradation, the porous PGS sample DSC curves begin to shift to the right as the material is degrading and returning back to the prepolymeric state. The greatest change is observed in the melting temperatures of the 120°C and 140°C as they increase through the weeks of degradation.

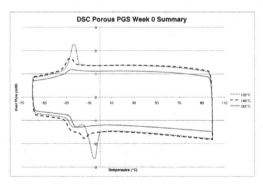

Figure 3. DSC week 0 curve summary of 120°C, 140°C, and 165°C cured porous PGS

Table I. Chart summary of the 120°C, 140°C, and 165°C DSC data for T_C, T_M, and T_g values.

PGS Porous 120°C Cure				PGS Porous 140°C Cure				PGS Porous 165°C Cure			
Week	T_C (°C)	T_M (°C)	T_g (°C)	Week	T_C (°C)	T_M (°C)	T_g (°C)	Week	T_C (°C)	T_M (°C)	T_g (°C)
0	-23.87	-4.61	-27.21	0	-27.64	-12.99	-29.22	0	-26.34	-21.99	-25.09
4	-25.58	-5.00	-30.19	4	-29.40	-23.29	-29.30	4	-29.39	-20.65	-29.36
8	-22.96	-3.23	-29.35	8	-26.28	-21.68	-25.43	8	-28.68	-23.16	-28.20
12	-27.68	-0.92	-27.03	12	-26.39	-22.03	-25.95	12	-26.58	-21.67	-25.47
16	-22.87	3.70	-21.73	16	-25.27	-4.04	-29.41	16	-26.31	-21.91	-25.47

TGA

The thermal stability of the porous PGS was examined through TGA and the summary curves for 120°C, 140°C, and 165°C cure temperatures are featured in Figure 2. High ending percentages from week 0 samples may indicate any residual salt from processing before samples were submerged in SBF. The onset temperatures during biodegradation, which signify the temperature where mass loss begins. During degradation, the onset temperatures increase and the curves shift right, closely matching the curve of the prepolymer.

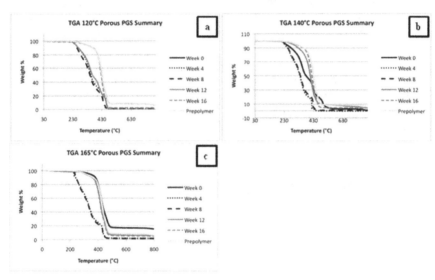

Figure 4. TGA curves of PGS cured at (a) 120°C, (b) 140°C, and (c) 165°C.

CONCLUSIONS

When PGS is cured at high cure temperatures, there is a greater crosslinking density and therefore less free chains are available to align in a crystalline structure. Degradation increases with lower cure temperature. Greatest degradation occurs in 120°C cure with pore collapse and return to the prepolymeric state, while comparatively less degradation occurs in 140°C and165°C.

REFERENCES

1. Wang Y, Kim YM, Langer R. (2002) *In vivo* degradation characteristics of poly(glycerol sebacate). J Biomed Mater Res 66A: 192–197.
2. Wang Y, et al. (2002) A Tough Biodegradable Elastomer. Nature biotechnology 20.6: 602-606.
3. Sun Z, Chen C, Sun M, Ai C, Lu X, et al. (2009) The application of poly (glycerol-sebacate) as biodegradable drug carrier. Biomaterials 30(28): 5209-5214.
4. Motlagh D, Yang J, Lui KY, Webb AR, Ameer GA. (2006) Hemocompatibility evaluation of poly(glycerol-sebacate) in vitro for vascular tissue engineering. Biomaterials 27(24): 4315-4324.
5. Gao J, Ensley AE, Nerem RM, Wang Y. (2007) Poly(glycerol sebacate) supports the proliferation and phenotypic protein expression of primary baboon vascular cells. J Biomed Mater Res Part A 83A(4): 1070-1075.
6. Gao J, Crapo P, Nerern R, Wang Y. (2008) Co-expression of elastin and collagen leads to highly compliant engineered blood vessels. J Biomed Mater Res Part A 85A(4): 1120-1128.10. Crapo PM, Wang Y. (2010) Physiologic compliance in engineered small-diameter arterial constructs based on an elastomeric substrate. Biomaterials 31(7): 1626-1635.
7. Sales VL, Engelmayr GC, Jr., Johnson JA, Jr., Gao J, Wang Y, et al. (2007) Protein precoating of elastomeric tissue-engineering scaffolds increased cellularity, enhanced extracellular matrix protein production, and differentially regulated the Radisic M, Park H, Chen F, Salazar-Lazzaro JE, Wang Y, et al. (2006) Biomirnetic approach to cardiac tissue engineering: Oxygen carriers and channeled scaffolds. Tissue Eng 12(8): 2077-2091.
8. Radisic M, Park H, Gerecht S, Cannizzaro C, Langer R, et al. (2007) Biomimetic approach to cardiac tissue engineering. Philos Trans R Soc B-Biol Sci 362(1484): 1357-1368.
9. Maidhof R, Marsano A, Lee EJ, Vunjak-Novakovic G. (2010) Perfusion seeding of channeled elastomeric scaffolds with myocytes and endothelial cells for cardiac tissue engineering. Biotechnol Prog 26(2): 565-572.
10. LeBlon CE, Pai R, Fodor CR, Golding AS, Coulter JP, Jedlicka SS. (2012). *In vitro* comparative biodegradation analysis of salt-leached porous polymer scaffolds.
11. Kokubo T, Kushitani H, Sakka S, Kitsugi T, Yamamuro T. (1990) Solutions able to reproduce *in vivo* surface-structure changes in bioactive glass-ceramic A-W^3.

Mater. Res. Soc. Symp. Proc. Vol. 1621 © 2014 Materials Research Society
DOI: 10.1557/opl.2014.359

Novel Absorbable Polyurethane Biomaterials and Scaffolds for Tissue Engineering

Syam P. Nukavarapu[1,2,3,5], Rao S. Bezwada[4], Deborah L. Dorcemus[1,2], Neeti Srivasthava[4], Robert J. Armentano[1]

[1]Institute for Regenerative Engineering, University of Connecticut Health Center Farmington, CT 06030, U.S.A
[2]Biomedical Engineering, University of Connecticut Storrs, CT 06269, U.S.A
[3]Materials Science & Engineering, University of Connecticut Storrs, CT 06269, U.S.A
[4]Bezwada Biochemical, LLC., Hillsborough, NJ, US.
[5]Orthopedic Surgery, University of Connecticut Health Center, Farmington, CT 06030, U.S.A

ABSTRACT

This study reports a novel class of biodegradable polyurethane biomaterials and three-dimensional scaffolds for tissue engineering. Solvent casted polyurethane films were studied for biocompatibility by seeding with human bone marrow derived stromal cells. In order to develop a three-dimensional and porous structure, a dynamic solvent sintering method was applied to the polyurethanes for the first time. Microstructural studies on the sintered scaffolds reveal porous structure formation with bonding between the adjacent microspheres. In conclusion, this study establishes new polyurethane biomaterials that are fully absorbable for tissue engineering applications.

INTRODUCTION

Polyurethanes (PUs) are an important class of biomaterials developed in the past primarily for biostable biomedical devices and their coatings [1-4]. Although polyurethanes find a wide range of industrial applications, segmented PU elastomers are of relevance to the medical industry due to their toughness, durability, biocompatibility, and biostability [2, 5]. In general, segmented polyurethanes are block co-polymers with an aliphatic polyol soft segment, and an aliphatic or aromatic hard segment; the soft segment is formed by the reaction between isocynate and polyol, while the hard segment is formed through a reaction between isocynate and chain extender. Segmented polyurethanes are prepared in two ways: (i) a single step process wherein isocynate, chain extender, and polyol are reacted together, or (ii) in a two step process wherein the isocynates and polyol are reacted to form pre-polymer, and then the pre-polymer is reacted with a chain extender [6, 7].

Recently, several segmented polyurethanes were synthesized to be biodegradable with the aim of developing biodegradable polyurethane scaffolds for tissue engineering and regenerative medicine applications. In this direction, several putative biodegradable polyurethanes are synthesized with polyester soft segments, and aliphatic or aromatic isocynate hard segments. These polyurethanes in various shapes and sizes have been investigated *in vitro* and *in vivo* for biocompatibility and biodegradability, and their ability to support cell in growth and tissue regeneration [8-12]. The main drawback of the existing class of polyurethanes is the lack of complete degradation. In the majority of cases, the soft segments of these polymers degrade but the hard segments, consisting of non-degradable isocyanates, remain hydrolytically non-degradable [13].

The aim of this study is to introduce a new class of polyurethanes that show hydrolytic degradation of not only the soft segments, but also the hard segments. As part of this study, two polyurethanes are synthesized and evaluated for biocompatibility, and fabricated into three-dimensional and porous structures as scaffolds for bone and osteochondral tissue engineering.

MATERIALS AND METHODS

Polymer Synthesis: A new family of polyurethanes was synthesized by reacting biodegradable diisocynates and polycaprolactone (PCL, MW 2000). 1,4 Butane diol (BDO) was used as a chain extender. In brief, Isocynate and (Polyol+BDO) with a 1 to (0.7883+0.2424) ratio were mixed and the target polymers were synthesized through a one step melt polymerization method [13]. The biodegradable isocynates that were used in this study are shown below. Polymer prepared with Diethylene glycol diglycolate isocynate is referred to Diglycolate-PU and the polymer with Diethylene glycol dilactate diisocynate is referred to Dilactate-PU.

Diethylene glycol diglycolate diisocyanate (Diglycolate)

Diethylene glycol dilactate diisocyanate (Dilactate)

Polymer Films: Polyurethane films were fabricated using a solvent casting method [14]. In this method, polymers were dissolved in methylene chloride by vortexing the mixture. The polymer solution was subsequently poured into a petri dish lined with Bytac paper, and then the solvent was allowed to evaporate slowly at 4 °C for 24 h. Finally, circular discs of 10 mm were bored from the films and these films used for cell compatibility studies.

Scaffold Fabrication: Diglycolate-PU microspheres were made using an oil-in-water emulsion method [15, 16]. In brief, the polymer was dissolved in methylene chloride (20% wt/volume), and then the homogeneous solution was poured as a continuous flow into a 1% polyvinyl alcohol (PVA) solution stirred at 250 rpm. The resulting microspheres were sieved to obtain 425–600μm size range and preserved in a desiccator until further use. The microspheres were chemically sintered using a solvent/non-solvent method [15]. Solvent and non-solvent components used are tetrahydrofuran (THF) and n-hexane, respectively. A mixture of THF and n-hexane was added to microspheres in a 1:1.3 (volume to weight ratio) and vortexed for 10 seconds before packing them in a mold to obtain a three-dimensional and porous Diglycolate-PU scaffold. Sintered scaffolds were used for osteogenic evaluation, while "polymer-hydrogel" matrices (i.e., scaffolds infused with PuraMatrix gel) were used for chondrogenic evaluation [17].

Cell Studies: Human bone marrow aspirate (BMA) was concentrated for MSCs using Magellan System from Arteriocyte, OH. The device is capable of separating BMA into three fractions: red blood cells, platelet-rich plasma, and platelet-poor plasma. Platelet-rich plasma containing monocytes was cultured to obtain the MSC population, and passage 3 cells were used in this investigation [18, 19]. Polyurethane films and scaffolds were sterilized by immersing in 70% ethanol for 30 min followed by UV radiation. Samples were placed in 24-well plates and seeded with hBMSCs on each sample. Polymer films with cells (30k/sample) were cultured in basal media (DMEM/F-12 + GlutaMAX, 10% FBS and 1% P/S) for 7, 14 and 21 days and the cell proliferation was determined using pico-green assay [17]. PU microsphere scaffolds seeded with 50k and 500k/scaffold were evaluated for osteogenesis and chondrogenesis, respectively. Cell seeded scaffolds were cultured in basal media supplemented with 10nM dexamethasone, 10mM β-Glycerophosphate, and 50 μg/mL Ascorbate-2-Phosphate for osteogenic differentiation, and in serum free basal media with ITS+, 100μL/mL Sodium Pyruvate, 50ng/mL L-Proline, 50μg/mL Ascorbate-2-Phosphate, 10^{-7} M dexamethasone, and 10ng/mL TGF β1 for chondrogenic differentiation. After 21 days of culture, the scaffolds were stained with Alizarin red and Alcian blue to confirm hMSC osteogenic and chondrogenic differentiation, respectively.

DISCUSSION

Diglycolate- and Dilctate-Polyurethanes: Hydrolytically degradable aromatic Diglycolate and Dilactate isocynates were synthesized. A single step reaction of these isocynates with PCL (MW 2000) (polyol) and 1,4 butane diol (chain extender) resulted in the synthesis of Diglycolate and Dilctate based polyurethanes. Unlike 4,4'-methylenebis(phenyl isocynate) (MDI) based polyurethanes, Diglycolate and Dilactate isocynates are comprised of biodegradable glycolide or lactide groups in place of the methylene linkage. Therefore, it is possible that the new polyurethanes with biodegradable hard segments are fully absorbable. The degradation rates of Dilactate-PU should be much slower than Diglycolate-PU, as the lactate group is hydrolytically more stable than the glycolate group. Expected degradation products are not toxic since PCL, PGA or PLLA, and PEG are key components of absorbable medial devices. PGA or PLLA degradation into acids is not a problem in this case, as the concentration of acid levels are considered to be very low. While the polymer degradation is not the focus of this study we characterized them for their suitability as biomaterials, and their processability into 3D-scaffolds for bone and osteochondral tissue engineering.

Cell Compatibility: Good cell compatibility is one of the key characteristics of a biomaterial for its consideration for tissue engineering applications. Here, we chose to study cell seeded polymer films and determine the amount of DNA present at culture periods 7, 14 and 21 days. Since the DNA amount is proportional to the number of cells on a matrix, this data in a way shows the number of cells on a matrix at a given time point. The DNA amounts recorded for Diglycolate-PU, Dilactate-PU and PLGA 85/15 are shown in Figure 1. PLGA 85/15 serves a control in this experiment. On day 7, polyurethane samples displayed significantly higher amounts of DNA compared to that of the control. The trend is similar even on day 14, but by day 21 the PUs showed DNA amounts somewhat similar to PLGA. Among the polyurethanes, Diglycolate-PU shows higher amounts of DNA at all time points when compared with the Dilactate-PU group. Unlike PLGA, polyurethanes do not show increase in cell number over the culture time. This could be attributed to a faster growth of the cells on the Diglycolate polyurethane matrices, however further experiments are needed to confirm this. The data obtained here suggest cell compatibility of the newly developed polyurethanes.

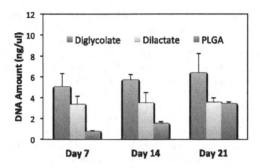

Figure 1: Cell proliferation assay for the Diglycolate and Dilactate polymers seeded with human bone marrow stromal cells, recorded at days 7,14, and 21.

Three-dimensional Scaffolds: Diglycolate polyurethane microspheres were brought together into a three-dimensional structure using a dynamic solvent sintering method. This method uses a solvent and non-solvent combination, which transition from a poor solvent to non-solvent over time [15]. The solvent and non-solvent in this case are chosen, as tetrahydrofuran (THF) and n-hexanes, respectively. A series of THF and n-hexanes compositions were tried to identify a suitable composition range for Diglycolate Polyurethane microsphere sintering. This study resulted in the identification of a range of compositions (THF to n-hexane volume ratio: 25 to 75 – 45 to 55) that support physically stable scaffold formation. SEM micrograph recorded on a typical composition (37.5 to 62.5) show a polyurethane scaffold with inter-connected pore structure (Figure 2a). The image recorded at higher magnification (shown in Figure 2b) confirms microsphere to microsphere bonding between the adjacent microspheres. This study shows the feasibility of processing polyurethanes into three-dimensional and porous scaffolds using a chemical sintering method.

Figure 2: SEM micrographs recorded on a solvent/non-solvent sintered Diglycolate polyurethane scaffold showing; (a) porous scaffold formation, and (b) bonding between the adjacent microspheres.

Scaffolds for Bone and Osteochondral Tissue Engineering: In addition to cell compatibility, a scaffold for bone and osteochondral tissue engineering should be evaluated for its ability to support MSC osteogenesis and/or chondrogenesis [20, 21]. For chondrogenesis experiments, the Diglycolate Polyurethane matrices were infiltrated with a PuraMatrix hydrogel containing hBMSCs. This is to provide optimal extracellular environment for MSCs to go through chondrogenic differentiation and cartilage matrix formation. As seen from Figure 3, Diglycolate polymer scaffolds cultured in osteogenic media and chondrogenic media display Alizarin red and Alcian blue staining, respectively. While PU scaffolds display weaker Alizarin red staining, advanced matrices display alcian blue staining comparable with the staining observed for the control PLGA group. We chose PLGA 3D scaffold in place of the conventional tissue culture plastic control due to its similarity in 3D structure. Overall, this preliminary study confirms the PU scaffold's osteogenic and chondrogenic ability and their potential use as scaffolds for bone and osteochondral tissue engineering.

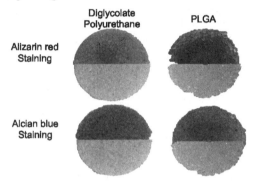

Figure 3: Three-dimensional scaffolds cultured for 21 days in osteogenic and chondrogenic media demonstrating positive staining of Alizarin red and Alcian blue for Diglycolate polyurethane and the control PLGA (top half stained and bottom half unstained).

CONCLUSIONS

Through this study, polyurethanes with isocynates comprising of hydrolytically degradable glycolide and lactide groups were developed as new potentially degradable polyurethane biomaterials. *In vitro* studies with human bone marrow derived stromal cells displayed biocompatibility of these new polymers. For the first time, polyurethane polymers were fabricated into three-dimensional and porous scaffolds using a liquid sintering method. Preliminary studies also indicate the fully absorbable polyurethane scaffold's osteogenic and chondrogenic nature, which suggests their potential as scaffolds for bone and osteochondral tissue engineering.

ACKNOWLEDGMENTS

Dr. Nukavarapu acknowledges partial support from AO Foundation (S-13-122N), NIH (AR062771), and NSF (1311907). He also acknowledges support from Connecticut Institute for Clinical and Translational Science (CICATS), the University of Connecticut. The authors also acknowledge support from the Raymond and Beverly Sackler Center for Biomedical, Biological, Physical and Engineering Sciences Center, and the Institute for Regenerative Engineering, at the University of Connecticut.

REFERENCES

1. Lelah MD, and Cooper JL. Polyurethanes in Medicine. Boca Raton, FL: CRC Press, 1987 (Biostability, Durability, Medical Applications).
2. Phillips RE, Smith MC, Thoma RJ. Biomedical applications of polyurethanes: implications of failure mechanisms. J Biomater Appl, 3 (1988), pp. 207–227.
3. Chawla AS, Blais P, Hinberg I, Johnson D. Degradation of explanted polyurethane cardiac pacing leads and of polyurethane. Biomater Artif Cells Artif Organs, 16 (1988), pp. 785–799
4. Benoit FM. Degradation of polyurethane foams used in the Meme breast implant. J Biomed Mater Res, 27 (1993), pp. 1341–1348.
5. Stokes K, McVenes R. Polyurethane elastomer biostability. J Biomater Appl 9, 321, 1995
6. Szycher M. Szycher's Handbook of Polyurethanes. Boca Raton, FL: CRC Press, 1999
7. Guelcher SA. Biodegradable polyurethanes: synthesis and applications in regenerative medicine. Tissue Eng Part B Rev. 2008 Mar;14(1):3-17.
8. Skarja GA, Woodhouse KA. Synthesis and characterization of degradable polyurethane elastomers containing an amino acid based chain extender. J Biomater Sci Polym Ed 9, 271, 1998.
9. Guan J, Sacks M, Beckman E, Wagner W. Biodegradable poly(ether ester urethane)urea elastomers based on poly(ether ester) triblock copolymers and putrescine: synthesis, characterization and cytocompatibility. Biomaterials 25, 85, 2003.
10. Cohn D, Hotovely-Salomon A. Designing biodegradable multiblock PCL/PLA thermoplastic elastomers. Biomaterials 26, 2297, 2005
11. Saad B, Hirt TD, Welti M, Uhlschmid GK, Neuenschwander P, Suter UW. Development of degradable polyesterurethanes for medical applications: in vitro and in vivo evaluations. J Biomed Mater Res 36, 65, 1997.
12. Borkenhagen M, Stoll RC, Neuenschwander P, Suter UW, Aebischer P. In vivo performance of a new biodegradable polyester urethane system used a nerve guidance channel. Biomaterials 19, 2155, 1998.
13. Bezwada RS. Absorbable Polyurethanes, ACS Symposium Series in Biomaterials, 2010.
14. Deng M, Nair LS, Nukavarapu SP, Kumbar SG, Jiang T, Krogman NR, Singh A, Allcock HR, Laurencin CT. Miscibility and in vitro osteocompatibility of biodegradable blends of poly[(ethyl alanato) (p-phenyl phenoxy) phosphazene] and poly(lactic acid-glycolic acid). Biomaterials. 2008 Jan;29(3):337-49.

15. Nukavarapu SP, Kumbar SG, Brown JL, Krogman NR, Weikel AL, Hindenlang MD, Nair LS, Allcock HR, Laurencin CT. Polyphosphazene/nano-hydroxyapatite composite microsphere scaffolds for bone tissue engineering. Biomacromolecules. 2008 Jul;9(7):1818-25.

16. Amini AR, Adams D, Laurencin CT, Nukavarapu SP. Optimally Porous and Biomechanically Compatible Scaffolds for Large Area Bone Regeneration. Tissue Eng Part A. 2012 Jul;18(13-14):1376-88.

17. Igwe J, Mikael P, Nukavarapu SP. Design, fabrication and in vitro evaluation of a novel polymer-hydrogel hybrid scaffold for bone tissue engineering. J Tissue Eng Regen Med. 2014; 8: 131–142.

18. Mikael P, Barnes B, Nukavarapu SP. Autologusly Enriched Human Bone Marrow Aspirate for Bone Tissue Engineering. Society for Biomaterials 2014 Annual Meeting, Denver, CO.

19. Mikael, P. E., and Nukavarapu, S. P., 2014, "Cell-Based Approaches for Bone Regeneration," in Bone Graft Substitutes, ASTM International (in press).

20. Amini AR, Laurencin CT, Nukavarapu SP. Bone Tissue Engineering: Recent Advances and Challenges. Crit Rev Biomed Eng. 2012, 40(5), 363-408.

21. Nukavarapu SP, Dorcemus D. Osteochondral Tissue Engineering: Current Strategies and Challenges. Biotechnology Advances, 31, 706-721, (2013).

Mater. Res. Soc. Symp. Proc. Vol. 1621 © 2014 Materials Research Society
DOI: 10.1557/opl.2014.346

Control of Cell Adhesion and Detachment on Temperature-Responsive Block Copolymer Langmuir Films

Morito Sakuma[1,2], Yoshikazu Kumashiro[2], Masamichi Nakayama[2], Nobuyuki Tanaka[2], Umemura Kazuo[1], Masayuki Yamato[2] and Teruo Okano[2]
[1]Department of Physics, Tokyo University of Science, 1-3 Kagurazaka, Shinjuku-ku, Tokyo, 162-8601 JAPAN
[2]Institute of Advanced Biomedical Engineering and Science, Tokyo Women's Medical University (TWIns), 8-1 Kawada-cho, Shinjuku-ku, Tokyo, 162-8666 JAPAN

ABSTRACT

This study used Langmuir-Schaefer (LS) method to produce thermo-responsive poly(N-isopropylacrylamide) (PIPAAm) modified surface. Block copolymer composed of polystyrene (PSt) and PIPAAm was synthesized by RAFT polymerization. PSt-*block*-PIPAAm (St-IP) with various chemical compositions was dropped on an air-water interface and formed Langmuir film by compression. Then, the Langmuir film changing a density was transferred on a hydrophobic modified glass substrate to produce St-IP transferred surface (St-IP LS surface). From the observation of atomic force microscope images, a nanostructure was observed on the transference of Langmuir films. Cell adhesion and detachment were also evaluated on the LS surfaces in response to temperature. Cell adhesion on LS surfaces at 37 °C was controlled by changing the chemical compositions and densities. After reducing temperature to 20 °C, adhering cells rapidly detached themselves with lower A_m and higher composition of PIPAAm. Our method should be proved novel insights for investigating cell adhesion and detachment on thermo-responsive surfaces.

INTRODUCTION

Poly(N-isopropylacrylamide) (PIPAAm) exhibits temperature-dependent behaviors above/below their lower critical solution temperature (LCST) at the vicinity of 32 °C in aqueous solutions [1-3]. This specific characteristic of PIPAAm molecule leads to the various applications on biomedical engineering. As one of the most famous applications, cell adhesion and detachment control on PIPAAm-grafted surface has been investigated intensively. Confluently cultured cells at 37 °C on the surface are harvested as intact sheet-like cells (so called "cell sheet") by lowering temperature below the LCST without enzymatic treatment [4,5]. For fabricating a cell sheet, temperature-responsive cell culture surfaces were prepared by various methods including grafting N-isopropylacrylamide monomer by electron beam and physical polymer coating methods such as spin coating [6-9]. This study developed a Langmuir-Schaefer (LS) method, as one of the physical polymer coating methods, for fabricating PIPAAm-deposited surfaces. The PIPAAm deposited surfaces with various area per molecule (A_m) and chemical compositions were characterized and cell adhesion and detachment characters on the surfaces were also discussed.

EXPERIMENT

Block copolymer of polystyrene-*block*-poly(*N*-isopropylacrylamide) (St-IP) with specific molecular weights were synthesized via reversible addition-fragmentation chain transfer radical (RAFT) polymerization. Synthesized St-IP was dissolved in chloroform and carefully spread onto a trough filled with deionized water. The polymers were squeezed with a pair of barriers and formed Langmuir film on an air-water interface. Surface pressure was measured with platinum Wilhelmy plate during squeezing the interface to evaluate the phase of the film. After forming the film, hydrophobic modified glass substrate was placed horizontally on the films. Then, the substrate was horizontally removed and temperature-responsive Langmuir Schaefer surfaces (St-IP LS surfaces) were obtained. The LS surfaces were analyzed by atomic force microscopy (AFM) (Veeco, Santa Barbara, CA, USA) for observing the transference of the Langmuir film, and cell adhesion and detachment in response to temperature on the surface was also evaluated.

Bovine carotid artery endothelial cells (BAECs) were incubated in TCPS dishes containing 10 mL of DMEM supplemented with 10% FBS, 100 U mL^{-1} penicillin, and 100 µg mL^{-1} streptomycin at 37 °C under a humidified atmosphere with 5% CO_2 and 95% air. After being confluently cultured on TCPS dishes, BAECs were recovered from the dishes with 0.25% trypsin-EDTA and then seeded on a LS surface at a concentration of 1.0×10^4 cells cm^{-2} in DMEM supplemented with 10% FBS. After being incubated at 37 °C and 5% CO_2, adhering BAECs were observed at 24 h after seeding by a phase contrast microscope (ECLIPSE TE2000-U) (Nikon, Tokyo). Then, for investigating a temperature effect on cell detachment on a LS surface, the cells were incubated at 20 °C at 5% CO_2 for 120 min, and the number of adhering cells was counted.

DISCUSSION

The chemical composition of obtained block copolymer (St-IP) was characterized by ^{1}H NMR. In this study, the number of St chains was defined at 130 and the number of IP chains was modulated by a reaction condition. Synthesized block copolymers were abbreviated as follows; St : IP=130 : 110 (St-IP110), 130 : 170 (St-IP170), 130 : 240 (St-IP240), 130 : 480 (St-IP480). Langmuir films with various densities, which was controlled area per molecule (A_m) evaluated by surface pressure measurement, were fabricated at an air-water interface. Figure 1 shows the surface pressure-area (π-A) isotherms of St-IP Langmuir films at an air-water interface. The values of surface pressure of Langmuir films were gradually increased and reached to a saturation state (30 mN m^{-1}). After the saturation state, the films reached to a condensed phase and moved to a collapsed phase by the further compression. The result of phase transition of St-IP on an air-water interface was similar on previous studies with a block copolymer composed of polystyrene and hydrophilic polymer [10]. Then, the films were transferred on hydrophobic modified glass substrates with A_m from 3 to 40 nm^2 molecule^{-1}.

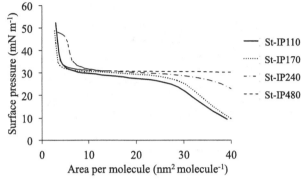

Figure 1. Surface pressure-area (π-A) isotherms of St-IP Langmuir films at an air-water interface. The y-axis shows the surface pressure of the film, and the x-axis shows the area per molecule.

Characterization of St-IP LS surface

Langmuir film transferred surfaces (St-IP LS surface) with A_m of 3 nm^2 molecule^{-1} and 40 nm^2 molecule^{-1} were characterized by AFM measurement (Figure 2). The LS surfaces showed nanostructures comparing for the bare hydrophobic substrate. In addition, since the amount of transferred St-IP was almost the same as that of Langmuir film at an air-water interface by the analysis of ATR FT-IR measurement, Langmuir films were successfully transferred on the hydrophobic substrates. The number of the nanostructures on the surface was depended on transferred density. At 40 nm^2 molecule^{-1} (Figure 2(B)), the LS surfaces showed cylindrical nanostructures, and the height and width of the nanostructures were found to be approximately 6-8 nm and 24-26 nm, respectively. The number of nanostructures on the LS surfaces was increased with increasing density from 40 to 3 nm^2 molecule^{-1}. In particular, the width of nanostructures on St-IP110 LS surface at 3 nm^2 molecules^{-1} was approximately 35 nm and the distance among nanostructures was approximately half of that on the LS surfaces at 40 nm^2 molecules^{-1}. Similar surface morphology dependencies were also observed on St-IP 170, St-IP240 and St-IP480 LS surfaces with various A_ms. These results indicated that the surface topography of the Langmuir films was controlled by changing the transferred density.

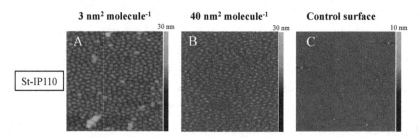

Figure 2. Atomic force microscope topographic images of St-IP LS surfaces. Image (A) shows the St-IP 110 LS surface having an area per molecule of 3 nm^2 molecule^{-1} and image (B) shows 40 nm^2 molecule^{-1}. Image (C) shows the hydrophobic modified glass substrate as a control. Scanning was examined by a X-Y tapping mode with phosphorous-doped silicon cantilever.

Cell adhesion and detachment on LS surface

The effects of density of St-IP and PIPAAm molecular weight toward temperature-dependent cell adhesion and detachment were investigated (Figure 3). On the LS surfaces at 3 nm^2 molecule^{-1}, the numbers of adhering cells were below 0.6 cells cm^{-2}, which was much lower density compared with the control surface. Notably, the numbers on St-IP110 LS surface at 3 nm^2 molecule^{-1} gradually decreased at 24 h after the incubation, indicating that the adhesion force between cells and the LS surfaces was not strong, followed by the spontaneous detachment. At 40 nm^2 molecule^{-1}, the number of adhering cells was 0.8-0.9 cells cm^{-2}, which was almost the same as the control surface. Figure 3 also shows the cell detachment on the LS surfaces after reducing temperature from 37 °C to 20 °C for 2 h. At 3 nm^2 molecule^{-1}, almost cells were detached from the LS surfaces by 30 min incubation. At 40 nm^2 molecule^{-1}, cells were gradually detached themselves from St-IP110, St-IP170 and St-IP240 LS surfaces, and perfectly detached from St-IP480 surfaces. These results indicated that (1) lower density was effective for increasing cell adhesion at 37 °C and (2) higher density and increasing PIPAAm molecular weight of St-IP molecule was dominant for acceleration cell detachment.

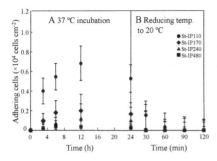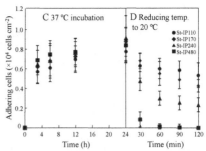

Figure 3. Cell adhesion and detachment assay on St-IP LS surfaces. The left graphs (A) and (B) show adhering cells ($\times 10^4$ cells cm^{-2}) on St-IP LS surfaces having an area per molecule of 3 nm^2 molecule^{-1} of St-IP110 (●), St-IP170 (♦), St-IP240 (▲), and St-IP480 (■). The right graphs (C) and (D) show adhering cells ($\times 10^4$ cells cm^{-2}) on St-IP LS surfaces having an area per molecule of 40 nm^2 molecule^{-1}. The graphs on the left and right columns show cell adhesion process at 37 °C and cell detachment process by reducing temperature from 37 °C to 20 °C for 120 min, respectively.

CONCLUSIONS

In this study, Langmuir films composed of polystyrene-*block*-poly(*N*-isopropylacrylamide) (St-IP) with various densities and chemical compositions were produced at an air-water interface, and the films were transferred on a hydrophobic modified glass substrate. Atomic force microscopy (AFM) topographic images showed nanostructure on the transferred Langmuir films. From the results of cell adhesion on the LS surfaces, the number of adhering cells was increased on the surface having lower density of St-IP. Cell detachment from the LS surfaces was accelerated with increasing density of St-IP after reducing temperature. Higher molecular weight of PIPAAm also accelerated cell detachment after reducing temperature, assuming that higher molecular weight of PIPAAm effectively and strongly hydrated by temperature change. The precise control of density and composition of block copolymer was important factor for modulating cell attachment and detachment in response to temperature. Our method could be applied to produce tailor-made cell culture substrates.

ACKNOWLEDGMENTS

This study was supported by Creation of innovation centers for advanced interdisciplinary research areas Program in the Project for Developing Innovation Systems "Cell Sheet Tissue Engineering Center (CSTEC)" from the Ministry of Education, Culture, Sports, Science and Technology (MEXT), Japan. The authors are grateful to Dr. N. Ueno (Tokyo Women's Medical University) for his valuable comments and suggestions.

REFERENCES

1. Y. Li and T. Tanaka, *J. Chem. Phys.* **92**, 1365 (1990).
2. M. Heskinsa and J. E. Guilleta, *J. Macro. Sci.* **2**, 1441 (1968).
3. Y. Maeda, T. Higuchi and I. Ikeda, *Langmuir* **16**, 7503 (2000).
4. J. Yang, M. Yamato, C. Kohno, A. Nishimoto, H. Sekine, F. Fukai and T. Okano, *Biomaterials* **26**, 6415 (2005).
5. M. Yamato and T.Okano, *Mater. Today* **7**, 42 (2004).
6. K. Fukumori, Y. Akiyama, Y. Kumashiro, J. Kobayashi, M. Yamato, K. Sakai, and T. Okano, *Macromol. Biosci.* **10**, 1117 (2010).
7. Y. Akiyama, A. Kikuchi, M. Yamato and T. Okano, *Langmuir* **20**, 5506 (2004).
8. M. Nakayama, N. Yamada, Y. Kumashiro, H. Kanazawa, M. Yamato and T. Okano, *Macromol. Biosci.* **12**, 751 (2012).
9. M. Sakuma, Y. Kumashiro, M. Nakayama, N. Tanaka, K. Umemura, M. Yamato and T. Okano, *J. Biomater. Sci., Polym. Ed.* **25**, 431 (2014).
10. S. Nagano, Y. Matsushita, Y. Ohnuma, S. Shinma and T. Seki, *Langmuir* **22**, 5233 (2006).

Mater. Res. Soc. Symp. Proc. Vol. 1621 © 2014 Materials Research Society
DOI: 10.1557/opl.2014.200

Biofunctional Thermo-Responsive Polymeric Surface with Micropatterns for Label Free Cell Separation

Yoshikazu Kumashiro[1], Jun Ishihara[1,2], Terumasa Umemoto[1], Kazuyoshi Itoga[1], Jun Kobayashi[1], Masayuki Yamato[1] and Teruo Okano[1]
[1] Institute of Advanced Biomedical Engineering and Science, Tokyo Women's Medical University,
8-1 Kawada-cho, Shinjuku-ku,Tokyo 162-8666, Japan
[2]Division of Cellular Therapy, The Institute of Medical Science, University of Tokyo,
4-6-1 Shirokanedai, Minato-ku, Tokyo 108-8039, Japan.

ABSTRACT

Thready stripe-patterned thermo-responsive surfaces were prepared and their surface properties were characterized. Prepared 3 μm wide stripe-patterned surfaces were evaluated by observing the adhesions and detachments of three types of cells: HeLa cells (HeLas), human umbilical vein endothelial cells (HUVECs), and NIH-3T3 cells (3T3s). Although cell adhesion and detachment in response to temperature were observed on all cells on a conventional thermo-responsive surface without patterns, the thermo-responsive surface with a 3 μm striped-pattern exhibited various cell adhesion properties. HeLas hardly adhered to the patterned surface even at 37 °C. On the other hand, although HUVECs adhered on the patterned surface at 12 h after incubation at 37 °C, the adhered HUVECs detached themselves after another 12 h incubation at 37 °C. 3T3s adhered to the patterned surface at 37 °C and detached themselves after reducing temperature to 20 °C. A mixture of HeLa, HUVEC and 3T3 was separated using their different specific cell-adhesion properties, and the composition of cells was analyzed by a flow-cytometry. As a result, the conventional thermo-responsive surface with a stripe-pattern was found to function as a cell-separating interface by using specific cell adhesion properties.

INTRODUCTION

Functional surfaces have attracted much attention through their induction of various potential applications, such as superhydrophobic and superhydrophilic surfaces inspiring the nanostructure of lotus leaf [1,2], stimuli-responsive surfaces in response to external stimuli (pH [3,4], temperature [5], ultra-violet light [6,7], and so on), and self-governing surfaces [8,9]. In our laboratory, thermo-responsive cell culture dishes having poly(N-isopropylacrylamide) (PIPAAm) covalently immobilized on tissue culture polystyrene, are used for preparing an artificial tissue consisting of two dimensional monolayer cells [10-13]. Cells adhere and proliferate at 37 °C on PIPAAm modified cell culture surface, because PIPAAms are hydrophobic owing to their dehydration. After reducing temperature to 20 °C, the adhered cells detach themselves as single cells or a contiguous cell sheet from the surfaces, due to the hydration and swelling of grafted PIPAAms on the culture surface.

Cell therapies including tissue engineering and cell injection have focused in the field of regenerative medicine that reproduces the lost functions of the tissue and organs, and some clinical trials by transplantation of cells have been already performed for recovering the lost function of patients [14]. For the transplantation of living cells, an effective cell separation and purification technology that can provide adequate purity, yield, and viability after a separation have been demanded. A cell separation technology that requires no modification on cell is preferable for allowing separated cells to be transplanted to the organs.

In this study, a thermo-responsive surface with a thready striped hydrophilic pattern, which was smaller in size than a cell and was larger than the length of focal adhesion, were designed and fabricated. Then, a stripe-patterned thermo-responsive surface was assumed to show different cell adhesions preserving cell detachment ability from the thermo-responsive surface. In addition, the effect of stripe-pattern on the thermo-responsive surface was also discussed for label-free cell separation by analyzing flow-cytometry.

EXPERIMENT

For evaluating the cell separation property of the prepared surface, firstly, HeLas, human umbilical vein endothelial cells (HUVECs), and NIH-3T3 cells were separately cultured on a conventional thermo-responsive surface as well as 3 μm-wide stripe-patterned PIPAAm surface at 37 °C for 24 h and allowed them to detach themselves at 20 °C for 2 h. After confirming the specific cell adhesion, the mixture of cell suspension (1.0×10^3 cells/cm^2 for each cell) was cultured on the patterned surface where the initial number of cells was 4.0×10^4 cells, and incubated for 12 h at 37 °C (figure 1). For recovering the cells, the following steps were used: (1) after the incubation, the medium of the cultured surface was recovered and exchanged to fresh medium twice at 37 °C, and then the cultured surface was incubated for another 12 h at 37 °C; (2) after the incubation for 24 h (12 and 12 h), the medium of the cultured surface was recovered and exchanged to fresh medium; (3) finally, the cultured surface was incubated at 20 °C for 2 h for recovering the released cells from the surface. Then, the recovered cells at 12 and 24 h after the start of incubation at 37 °C, and at 2h after the start of incubation at 20 °C were stained with 5 μg/mL PE-stained MHC class I chain-related gene A and B (MICA/MICB), Alexa488-stained CD31, and APC-stained CD140β at a concentration of 1.0×10^3 cells/$\square$L in PBS containing 2% BSA, respectively. HeLas, HUVECs, and 3T3s were specifically modified PE-stained MICA/MICB, Alexa488-stained CD31, and APC-stained CD140β, respectively.

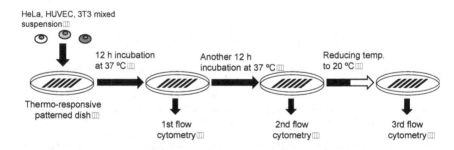

Figure 1. Schematic illustration of a cell separation system using 3 μm wide stripe-patterned thermo-responsive surface by exchanging medium and reducing temperature.

DISCUSSION

HeLas, HUVECs, and 3T3 cells were separately cultured on a conventional thermo-responsive surface as well as 3 μm-wide stripe-patterned PIPAAm surface at 37 °C for 24 h and allowed them to detach themselves at 20 °C for 2 h. On the thermo-responsive surface with a 3 μm stripe-pattern, the three different cells showed their different cell adhesion properties (figure 2). HeLas hardly adhered on the 3 μm stripe-pattern even at 37 °C. On the other hand, although HUVECs adhered on the patterned surface for 12 h at 37 °C, the adhered HUVECs spontaneously detached themselves after another 12 h incubation at 37 °C. 3T3s adhered to the patterned surface, and the adhered 3T3s detached themselves after reducing temperature to 20 °C not only from the patterned surface but also from a PIPAAm surface without a stripe-pattern. Since cell-surface interaction mediates multi protein-surface interactions including the interaction between extra cellular matrix and surface, the geometry of the proteins on cell surfaces should be different from different cell species, indicating that surface modification may be suitable for controlling cell adhesion.

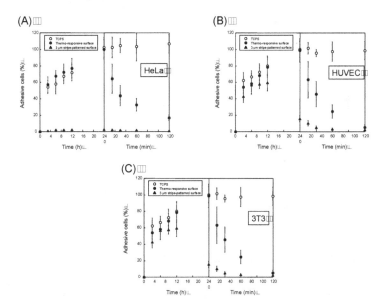

Figure 2. Time course of cell adhesion and detachment in response to temperature. HeLas (A), HUVECs (B), and 3T3s (C) were separately cultured on tissue culture polystyrene (the open circles), thermo-responsive surface (the closed circles), and 3 μm stripe-patterned surface (the closed triangles). Cells were incubated at 37 °C for 24 h (the left side of graph) and then at 20 °C for 120 min (the right side of graph).

Figure 3 shows the number of recovered cell at the defined time. From the medium obtained at 12 h, HeLa cells were mainly detected, and the ratio of HeLas, HUVECs, and 3T3s was approximately 73.3, 17.0, and 9.6%, respectively. Slightly higher ratio of HUVECs indicated that the initial adhesion of HUVECs on the patterned surface for 12 h was lower than those of controlled PIPAAm surface and tissue culture polystyrene (TCPS). From the medium obtained at 24 h, HUVECs were mainly detected, and the ratio of HeLas, HUVECs, and 3T3s was approximately 5.8, 82.4, and 11.8%, respectively. Finally, 3T3s were mainly detected from the medium obtained at 2h after 20 °C incubation, and the ratio of HeLas, HUVECs, and 3T3s was approximately 0.4, 6.1, and 93.5%, respectively. In addition, the amounts of recovered HeLas, HUVECs, and 3T3s were counted to be approximately 4.0×10^4 cells, 3.9×10^4 cells, and 7.6×10^4 cells, respectively. Considering the initial amount of cultured cell (4.0×10^4 cells on the surface), 3T3s were speculated to proliferate during the incubation at 37 °C for 24 h, and the amount of 3T3s was increased. Therefore, the designed stripe-patterned thermo-responsive surface was able to separate cells by the different adhesion properties of cells, preserving the cell detachment properties.

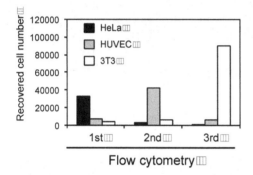

Figure 3. The number of recovered cells using the cell separation system of figure 1 analyzed by flow cytometry. The initial number of cultured cell was 4.0×10^4 cells. The area of the stripe-patterned surface was approximately 2.5 cm^2, and a total of 16 surfaces were used.

Preparation and characterization of stripe-patterned thermo-responsive dish

Thermo-responsive surface with thready striped PAAm patterns was prepared by similar method of previous study [15,16]. Photoirradiation was performed on g-line positive photoresist coated thermo-responsive culture dishes by a visible light through a quartz mask (3 μm black/white repeated pattern), followed by the development using a developer solvent. 3.5 mL of 20 wt% acrylamide (AAm) aqueous solution containing 5 wt% water soluble initiator was added on the photoresist-patterned culture dish, followed by radical polymerization on the developed position for 2 h by a blue light emitting diode (LED) (MIL-C1000T, MIL-U200, MILB18(A), SANYO Electric, Osaka, Japan). After the reaction, the dish was thoroughly washed with warm water and was immersed in the developer solvent again to remove the residual photoresist.

For the characterization of the thermo-responsive culture dish with striped PAAm-patterns, the patterned surface was stained by a fluorescent-protein. Alexa488 bovine serum albmin (BSA) (1:200 diluted) was dissolved in 10% PBS at a concentration of 10 □g/mL. Two mL of the protein solution was immersed to the patterned surface and the surface was incubated at 37 °C for 2 h. The patterned surface was then rinsed with PBS and examined under a fluorescence microscope. The depth of the patterned surface was obtained by the height mode of X-Y scanning tapping mode AFM (NanoScope® V Multimode) using a n-doped Si tip at room temperature. RMS values of the flattened images were further obtained using software.

Cell culture on stripe-patterned thermo-responsive surface

HeLas and 3T3s were cultured in DMEM supplemented with 10% FBS, 100 units/mL penicillin and 100 µg/mL streptomycin. The cells were recovered from tissue culture poly(styrene) (TCPS) dishes by treatment with 0.05% trypsin and 0.05% ethylenediaminetetraacetic acid (EDTA) in PBS and were routinely cultured passage at half confluency. Then, the cell suspension was cultured onto prepared stripe-patterned surfaces. For preparing cell suspension of HUVECs, HUVECs were cultured on a TCPS dishes with endothelial cell medium (EGM-2). HUVECs were recovered from the TCPS dishes by treating with 0.1% trypsin containing 1.1 mmol/L EDTA in HEPES buffered saline. The suspension of recovered HUVECs was fixed DMEM supplemented with 10% FBS, 100 units/mL penicillin and 100 µg/mL streptomycin, and then the HUVECs were seeded on treated stripe-patterned surfaces. The seeded density of the cells were fixed at 1.0×10^4 cells/cm^2 for single cells experiment. Adhesion of cultured cells on the patterned surfaces was monitored under a phase contrast microscope (Eclipse TE2000, Nikon, Tokyo). Then, for investigating a temperature effect on cell detachment from the patterned surfaces, the patterned surfaces were incubated at 20 °C and 5% CO_2 for 2 h, and the number of adhering cells was counted until 2 h for evaluation of cell detachment. For the cell separation experiment, the concentration of cell suspension was fixed at 1.0×10^3 cells/cm^2 for the experiment of cell separation

CONCLUSIONS

We prepared a stripe-patterned thermo-responsive surface and evaluated the surface for label-free cell separation. As the future aspects, a patterned surface would be used for separating differentiated cells and non-differentiated cell derived during the differentiation of stem cells. The patterned surface developed in this study has a potential for breakthrough in the next generation of biomaterials.

ACKNOWLEDGMENTS

This study was supported by Creation of innovation centers for advanced interdisciplinary research areas Program in the Project for Developing Innovation Systems "Cell Sheet Tissue Engineering Center (CSTEC)" from the Ministry of Education, Culture, Sports, Science and Technology (MEXT), Japan. The authors are grateful to Dr. N. Ueno (Tokyo Women's Medical University) for his valuable comments and suggestions.

REFERENCES

1. Z.-Z. Gu, H. Uetsuka, K. Takahashi, R. Nakajima, H. Onishi, A. Fujishima, and O. Sato, *Angew. Chem. Int. Ed.* **42**, 894 (2003).
2. N. A. Patankar, *Langmuir* **20**, 8209 (2004).
3. X. Yu, Z. Wang, Y. Jiang, F. Shi, and X. Zhang, *Adv. Mater.* **17**, 1289 (2005).
4. L. Tauk, A. P. Schröder, G. Decher, and N. Giuseppone, *Nat Chem* **1**, 649 (2009).
5. L. Chen, M. Liu, H. Bai, P. Chen, F. Xia, D. Han, and L. Jiang, *J. Am. Chem. Soc.* **131**, 10467 (2009).
6. Y. Jiang, P. Wan, M. Smet, Z. Wang, and X. Zhang, *Adv. Mater.* **20**, 1972 (2008).
7. H. Ge, G. Wang, Y. He, X. Wang, Y. Song, L. Jiang, and D. Zhu, *ChemPhysChem* **7**, 575 (2006).
8. T. Masuda, M. Hidaka, Y. Murase, A. M. Akimoto, K. Nagase, T. Okano, and R. Yoshida, *Angew. Chem. Int. Ed.* **52**, 7468 (2013).
9. U. Diebold, *Surf. Sci. Rep.* **48**, 53 (2003).
10. M. Yamato, and T. Okano, *Mater. Today* **7**, 42 (2004).
11. A. Kikuchi, and T. Okano, *J. Controlled Release* **101**, 69 (2005).
12. Y. Kumashiro, M. Yamato, and T. Okano, *Ann. Biomed. Eng.* **38**, 1977 (2010).
13. Y. Akiyama, A. Kikuchi, M. Yamato, and T. Okano, *Langmuir* **20**, 5506 (2004).
14. K. Nishida, M. Yamato, Y. Hayashida, K. Watanabe, K. Yamamoto, E. Adachi, S. Nagai, A. Kikuchi, N. Maeda, H. Watanabe, T. Okano, and Y. Tano, *N. Engl. J. Med.* **351**, 1187 (2004).
15. Y. Kumashiro, K. Itoga, Y. Kinoshita, M. Yamato, and T. Okano, *Chem. Lett.* **42**, 741 (2013).
16. Y. Kumashiro, T. Matsunaga, M. Muraoka, N. Tanaka, K. Itoga, J. Kobayashi, Y. Tomiyama, M. Kuroda, T. Shimizu, I. Hashimoto, K. Umemura, M. Yamato, and T. Okano, *J. Biomed. Mater. Res. A*, (2013) in press.

Mater. Res. Soc. Symp. Proc. Vol. 1621 © 2014 Materials Research Society
DOI: 10.1557/opl.2014.69

Preliminary Investigation of a Sacrificial Process for Fabrication of Polymer Membranes with Sub-Micron Thickness

Luke A. Beardslee[1], Dimitrius A. Khaladj[1], Magnus Bergkvist[1]
[1]SUNY College of Nanoscale Science & Engineering, Albany NY.

ABSTRACT

Here we present a single mask sacrificial molding process that allows ultrathin 2-dimensional membranes to be fabricated using biocompatible polymeric materials. For initial investigations, polycaprolactone (PCL) was chosen as a model material. The process is capable of creating 250-500 nm thin, through-hole PCL membranes with various geometries, pore-sizes and spatial features approaching 2.5 micrometers using contact photolithography. The technique uses a mold created from two layers of lift-off resist (LOR). The upper layer is patterned, while the lower layer acts as a sacrificial release layer for the polymer membrane. For mold fabrication, photoresist on top of the layers of lift-off resist is patterned using conventional photolithography. During development the mask pattern is transferred onto the first LOR layer and the photoresist is removed using acetone, leaving behind a thin mold. The mold is filled with a solution of the desired polymer. Subsequently, both the patterned and lower LOR layers are dissolved by immersion in an alkaline solution. The membrane can be mounted onto support structures pre-release to facilitate handling.

INTRODUCTION

Scaffold structures comprised of extracellular matrix (ECM) proteins provide cellular support and function within biological systems, particularly as basement layers for epithelial and endothelial cell sheets [1, 2]. They are often highly permeable membrane layers with micrometer to nanometer thickness [3, 4], where the Descemet's membrane within the cornea of the eye, and the basement membrane of the vascular endothelium are two examples. Proper formation of such basement membranes during *in-vitro* cell culture can help to facilitate growth of functional artificial tissue from primary or stem cell cultures. To support *in-vitro* growth of functional tissue and to promote basement membrane formation, a thin pore containing scaffold material with suitable geometries is desirable. In this work, we present lithographically fabricated molds for the creation of ultrathin micropatterned scaffolds for tissue engineering. Many different methods are possible for the creation of patterned/structured biomaterials systems, including electrospinning [5], direct photopatterning of hydrogels [6, 7], photocrosslinking of polymers [8-10], soft lithography [11-13], stamping [14], etching [15], UV crosslinking of hydrogels through a PDMS mold [16], and two photon polymerization [17].

Direct photopatterning is an attractive and a relatively straight forward approach that can work well. For example, radical crosslinking of hydrogels [6] and bioplastics [9] has been demonstrated. However, such techniques typically require synthesis of unique precursor molecules. A common alternative method to create textured polymer structures is soft lithography [6, 8-10, 17]. This approach is commonly used for materials that cannot be directly photopatterned; however the demolding approach can cause thin polymer structures to tear and/or deform, which limits the minimum thickness of the materials (typically 5-15 μm) [11, 12]. Also the fabrication of membranes with through-pores presents a practical challenge.

We are interested in techniques to create micropatterned membranes with submicron thickness that are (1) easy to release from a substrate for further manipulation (2) generally applicable to a variety of commonly used biomaterials. To realize these two goals, a microfabrication process is demonstrated that utilizes a sacrificial molding technique to produce freestanding, custom designed, polymer membranes with through-pores.

Sacrificial processes are common in MEMS (microelectromechanical systems) fabrication and allow for the release of ultrathin freestanding structures that are first fabricated on a solid substrate. In the process presented here a sacrificial structure is created that can be dissolved completely in photoresist developer (alkaline solution) after coating with a polymer. The sacrificial mold is made from lift-off resist (LOR). These resists are traditionally used for undercutting positive resist structures for subsequent metal evaporation. A variety of other water soluble sacrificial layers have been investigated [18], but none provide the well-defined dissolution rates of LOR. The predictable undercut rate of the LOR beneath the positive resist is one advantageous property that allows for its use as a molding material for defined geometries.

The undercut rate of the LOR is determined by the bake temperature after spin coating. Lift-off resists have several other advantageous properties in addition to a predictable undercut rate. They are soluble in aqueous photoresist developers (tetramethylammonium hydroxide (TMAH), or even NaOH), where a pattern defined in a positive resist can be transferred during development. LOR is insoluble in solvents typically used for biodegradable polymers (toluene, acetone etc.). The insolubility of LOR in these solvents is a key property that allows the molds to be coated without dissolving. For example, if photoresist on top of a single LOR layer was used as the mold, the release step in TMAH could be realized, but coating the mold with a biodegradable polymer would be problematic as the resist mold would dissolve in commonly used solvents. Also, two layers of LOR coated on top of one another and baked at different temperatures allow different dissolution rates between the two layers facilitating pattern transfer.

EXPERIMENT

In the iteration of the process presented here, SC1813 photoresist is used along with either LOR 3A or LOR 10A (Microchem, Newton MA) as the patterned LOR layer. The patterned layer of LOR is baked at 150 °C. The lower sacrificial layer is PMGI SF9 lift-off resist (Microchem, Newton MA). The PMGI is baked at 200 °C, and dissolves approximately 5 times slower than the LOR layer. After both of the lift-off resist layers have been coated, the SC1813 resist is spin-coated, baked, and exposed to transfer the mask pattern into the resist. During development of the SC1813, the mask pattern is transferred into the LOR layer. The insolubility of LOR in common organic solvents allows stripping of the photoresist with acetone leaving the underlying LOR mold. Subsequently, a spin coating step is used to fill the mold with the desired scaffold polymer. PCL and poly(glycidylmethacrylate) (PGMA) were chosen as polymers for the present work because they both have been investigated and shown promise as biomaterials for tissue engineering applications [8, 12, 19]. An overview of the fabrication process is given in Figure 1.

For molding PCL membranes a 0.025 g/ml PCL in toluene solution was spin coated onto the molds. For the PGMA membranes, molds were coated with a 0.005 g/ml PGMA solution in tetrahydrofuran. The molds were coated at spin speeds between 1500-3000 rpm. Before release the PGMA was crosslinked by a 60 min incubation in an ammonia atmosphere at 60 °C. After the polymer solution is spin coated onto the mold, the polymer membranes are released by placing the molds in 0.26M TMAH. The release typically takes 10-30 min. The depth of the mold can be controlled by making the upper LOR layer thicker and by developing the pattern for longer amounts of time. This iteration of the process is limited by the isotropic nature of the pattern transfer into the LOR as dissolution rates are similar in all directions. Thus, thicker membrane molds can be produced but at the cost of lower pattern resolution.

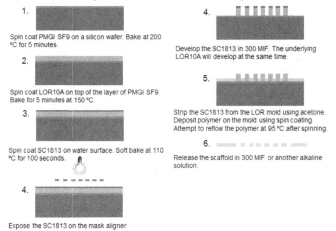

1. Spin coat PMGI SF9 on a silicon wafer. Bake at 200 °C for 5 minutes.

2. Spin coat LOR10A on top of the layer of PMGI SF9. Bake for 5 minutes at 150 °C.

3. Spin coat SC1813 on wafer surface. Soft bake at 110 °C for 100 seconds.

4. Expose the SC1813 on the mask aligner.

4. Develop the SC1813 in 300 MIF. The underlying LOR10A will develop at the same time.

5. Strip the SC1813 from the LOR mold using acetone. Deposit polymer on the mold using spin coating. Attempt to reflow the polymer at 95 °C after spinning.

6. Release the scaffold in 300 MIF or another alkaline solution.

Figure 1: Cross-sections giving an overview of the molding process. This process has been used to create PGMA and PCL membranes with a thickness of 250-500 nm.

RESULTS

Figure 2 shows light microscope images of the LOR molds after the SC1813 has been removed, but before they are coated with polymer. The feature sizes on the masks that were used to make the molds are between 2 and 3 micrometers. Additional work has been performed where the LOR layers can be reactive ion etched through a metal mask allowing for a more anisotropic profile and higher aspect ratio structures. Images of the released scaffolds are shown in Figure 3.

Figure 2: Light microscope images of the LOR 10A molds before polymer coating. The mold on the left is a hexagonal pattern and the mold on the right is a grid pattern. The scale bar on the lower left of each image measures 20 µm's.

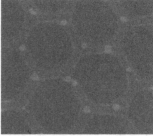

Figure 3: Optical microscope image of a PCL membrane molded in a hexagonal pattern (left). Environmental scanning electron microscopy (SEM) image of a hexagonal PGMA membrane (right). The hexagons in the images are laid out to have a vertex to vertex distance of 12 μm's. The width of each beam is around 3 μm's.

For any engineering application the membranes would need to be manipulated and freely moveable so that they can be inserted into different cell culture solutions or bioreactors. The membrane in Figure 4 is mounted to a Ø6.5 mm transwell fixture (Corning) to facilitate easy handling. This can be accomplished by applying a small amount of SU8 photoresist (Microchem, Newton MA) or PCL (heated to 95 °C) to the transwell fixture and then applying it to the mold. In the case of SU8, a bake is performed at 95 °C to remove the solvent from the resist and to harden it. With PCL, a bake at 95 °C is performed to re-flow the polymer to form a bond between the membrane and fixture after cooling to room temperature. In both cases the substrate with mold and transwell fixture can be placed in TMAH for release. When using the transwell fixture, the release time is greatly increased (between 1-16 hours) because of the slow diffusion of the TMAH beneath the several millimeter wide rim of the fixture.

An alternative mounting strategy to allow easier scaffold handling, and faster release times is to mount the structures onto an epoxy ring (Figure 4). In this method an SU8 ring (with either a 200 μm or 500 μm wide rim) is attached to the mold after coating with polymer. The ring is secured to the membrane using PCL as described for the transwell fixtures. The SU8 rings are fabricated from SU8 2010 (Microchem, Newton MA) and are approximately 25 μm's thick. The SU8 rings are fabricated on top of an LOR layer and released in 0.26M TMAH. The ring attachment allows thin polymer membranes to be released directly into solution and enable easy manipulation. The release time is typically 2-3 hours with the SU8 ring attached to the scaffold.

Figure 4: PCL membrane mounted on a transwell fixture (left). Three different hexagonal PCL scaffolds that have been mounted to PCL-coated SU8 rings and released (right). A US quarter is shown for a size comparison (right).

The released membranes range from 250 - 500 nm thick, as observed in SEM cross sections and by profilometer (data not shown). Figure 5 shows an image where a released membrane is folded over an edge, which allow both the top and the side of the scaffold to be visualized. As can be seen the thickness of the membrane is significantly less than the 3 micrometer scaffold beams.

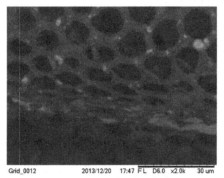

Grid_0012   2013/12/20   17:47 FL   D6.0 x2.0k   30 um

Figure 5: A view of a released PCL membrane folded such that a cross section of the beams can be seen. The beam width of the hexagons is 3 μm's. Comparing the beam width to the scaffold thickness, one can clearly see that the thickness is submicron.

The results presented above demonstrate that this sacrificial molding process is suitable for fabrication of sub-micron thin scaffolds with through-pores. The ultimate goal is to create scaffolds that enable cell growth, self-assembly of basement membrane, and establishment of functional endo/epithelial cell layers. Two initial experiments were performed to demonstrate scaffold integrity after extended exposure to physiological cell culture conditions. In the first, a PCL scaffold mounted to a transwell fixture was incubated at 37° C in 1X phosphate buffered saline (PBS) for 2 weeks. In a second experiment a PCL scaffold was used to culture trabecular meshwork cells similar to [20]. These experiments demonstrated that these thin membranes could survive long-term cell culture. Future work will include cell culture analysis on these scaffolds.

CONCLUSIONS

A sacrificial molding process has been demonstrated that can be used to produce sub-micron thin membranes using a variety of polymeric materials. In addition to simple molding, the polymers can be modified with gas-phase reactants after coating on the mold. This allows for both greater mechanical/chemical stability and modification of the polymers if needed for other reasons (such as chemical conjugation of adhesion proteins). The scaffolds made using the demonstrated process are constructed from widely used biomaterials, and many other common polymeric biomaterials could be used in this process. Future work will include cell culture on the scaffolds to further characterize their biocompatibility and biomaterials properties.

ACKNOWLEDGEMENT

The authors would like to acknowledge funding from the 2012 Collaboration Fund awarded by the Research Foundation of the State University of New York. The authors would also like to thank the research support staff at CNSE for help with equipment as well as Karen Torrejon for assistance with cell culture.

REFERENCES

1. P. D. Yurchenco, Cold Spring Harbor Perspectives in Biology, 3, a004911, (2012).
2. V. S. LeBleu, B. MacDonald and R. Kalluri, Experimental Biology and Medicine, 232, 1121, (2007).
3. A. L. James, P. S. Maxwell, G. Pearce-Pinto, J. G. Elliot and N. G. Carroll, American Journal of Respiratory and Critical Care Medicine, 166, 1590, (2002).
4. T. Osawa, M. Onodera, X. Y. Feng and Y. Nozaka, Journal of Electron Microscopy, 52, 435, (2003).
5. S. G. Kumbar, R. James, S. P. Nukavarapu and C. T. Laurencin, Biomedical Materials, 3, 034002, (2008).
6. W. Xiao, J. He, J. W. Nichol, L. Wang, C. B. Hutson, B. Wang, Y. Du, H. Fan and A. Khademhosseini, Acta Biomaterialia, 7, 2384, (2011).
7. K. E. Schlichting, T. M. Copeland-Johnson, M. Goodman, R. J. Lipert, T. Prozorov, X. Liu, T. O. McKinley, Z. Lin, J. A. Martin and S. K. Mallapragada, Acta Biomaterialia, 7, 3094, (2011).
8. A. Hayek, Y. Xu, T. Okada, S. Barlow, X. Zhu, J. H. Moon, S. R. Marder and S. Yang, Journal of Materials Chemistry, 18, 3316, (2008).
9. H. Kweon, M. K. Yoo, I. K. Park, T. H. Kim, H. C. Lee, H.-S. Lee, J.-S. Oh, T. Akaike and C.-S. Cho, Biomaterials, 24, 801, (2003).
10. C.-C. Lin, A. Raza and H. Shih, Biomaterials, 32, 9685, (2011).
11. W. L. Neeley, S. Redenti, H. Klassen, S. Tao, T. Desai, M. J. Young and R. Langer, Biomaterials, 29, 418, (2008).
12. S. Sodha, K. Wall, S. Redenti, H. Klassen, M. J. Young and S. L. Tao, Journal of Biomaterials Science, Polymer Edition, 22, 443, (2011).
13. G. Vozzi, C. J. Flaim, F. Bianchi, A. Ahluwalia and S. Bhatia, Materials Science and Engineering: C, 20, 43, (2002).
14. J. Nagstrup, S. Keller, K. Almdal and A. Boisen, Microelectronic Engineering, 88, 2342, (2011).
15. G. Shayan, N. Felix, Y. Cho, M. Chatzichristidi, M. L. Shuler, C. K. Ober and K. H. Lee, Tissue Engineering: Part C, 18, 667, (2012).
16. H.-C. Moeller, M. K. Mian, S. Shrivastava, B. G. Chung and A. Khademhosseini, Biomaterials, 29, 752, (2008).
17. F. Claeyssens, E. A. Hasan, A. Gaidukeviciute, D. S. Achilleos, A. Ranella, C. Reinhardt, A. Ovsianikov, X. Shizhou, C. Fotakis, M. Vamvakaki, B. N. Chichkov and M. Farsari, Langmuir, 25, 3219, (2009).
18. V. Linder, B. D. Gates, D. Ryan, B. A. Parviz and G. M. Whitesides, Small, 1, 730, (2005).
19. K. M. Ainslie and T. A. Desai, Lab on a Chip, 8, 1864, (2008).
20. K. Y. Torrejon, D. Pu, M. Bergkvist, J. Danias, S. T. Sharfstein and Y. Xie, Biotechnology and Bioengineering, 110, 3205, (2013).

Mater. Res. Soc. Symp. Proc. Vol. 1621 © 2014 Materials Research Society
DOI: 10.1557/opl.2014.360

Fabrication and Morphological Investigation of Multi-walled Electrospun Polymeric Nanofibers

Jamal Seyyed Monfared Zanjani[1], Burcu Saner Okan[2], Mehmet Yildiz[1] and Yusuf Menceloglu[1]

[1]Faculty of Engineering and Natural Sciences, Sabanci University, Tuzla, Istanbul 34956, Turkey
[2]Sabanci University Nanotechnology Research and Application Center, SUNUM, Tuzla, Istanbul 34956, Turkey

ABSTRACT

Multi-walled nanofibers with their outstanding properties have found expanding applications on drug delivery systems, biosensors, self-healing materials and many other state-of-the-art technologies. This work investigates the fabrication and morphological control of multi-walled structured electrospun polymeric nanofibers by multi-axial electrospinning system. This process is based on a nozzle allowing multi-axial extrusion of different fluids with concentric orders. Two spinnable polymers of poly(methyl methacrylate) and polyacrylamide are chosen for the fabrication of middle and outer walls of co-axial hollow nanofibers, respectively. Hansen's solubility parameters are used to systematically optimize the solvent selection for each layer and control the degree of miscibility of layers with the purpose of tailoring the final wall morphology of nanofibers. Characterization studies are performed by Scanning Electron Microscopy, Energy-Dispersive X-ray Spectroscopy, Fourier Transform Infrared Spectroscopy, and Thermal Gravimetric Analyzer.

INTRODUCTION

Electrospinning is a promising, versatile, single-step and efficient technique to fabricate multi-walled structured nanofibers with a controllable diameter, wall thickness, mechanical properties, and surface functionalities [1]. The fabrication process of multi-axial electrospun nanofibers is based on a nozzle containing concentric tubes allowing for the extrusion of different fluids to tip of the nozzle under high voltage power. Bending instabilities and whipping motions applied on polymeric jet in electric field between nozzle and collector result in the reduction of the jet diameter and the formation of fibers with diameter ranging from several nanometers to micron. The layered structure and surface morphology of electrospun nanofibers are controlled by the degree of miscibility of solutions in each layer, polymer solution concentration, solvent vapor pressure, applied voltage, electrospinning distance, and flow rate [2]. In multi-axial electrospinning technique, the outer shell solution can be a polymeric materials with viscoelastic properties whereas the core solution can be either viscoelastic or Newtonian liquids [3]. The utilization of electrospinning as an encapsulation technique benefits from being a simple one-step, and continuous process in comparison to the chemically complex and expensive encapsulation methods.

Joo et al. [4] combined the advantages of nanofiber mats with the self-assembly functionality of block-copolymers as the intermediate layer and silica as core and shell layers by tri-axial electrospinning. Furthermore, Chen et al. [5] utilized tri-axial electrospinning to develop nanowire-in-microtube nanofibers and obtained homogeneous or heterogeneous wire-in-tube one dimensional materials by removing the middle spacer layer. In addition, Liu et al. [6] fabricated gelatin/poly(ε-caprolactone)/gelatin multi-layer nanofibers via tri-axial electrospinning technique to enhance mechanical properties of gelatin as a natural biodegradable and biocompatible polymer by a layer of synthetic poly(ε-caprolactone) with fine mechanical properties. Tri-axial electrospun nanofibers were also designed as novel drug delivery systems by the combination of multiple drug molecules in core and shell of the fibers with different release time profiles [7].

In present work, multi-axial electrospinning technique is applied to fabricate novel architecture of co-axial hollow fibers by using different middle and outer wall materials. Hansen's solubility parameters are used to control hollowness of the fibers and define the ideal solvent systems for wall solutions. The surface morphologies and wall diffusion are monitored by changing solvent systems. The interactions between middle wall and outer walls solutions are investigated by tailoring the solvent system. This study carries a significant importance to produce hollow fibers covered by two distinct walls. The distinction of these walls is adjusted by using binary solvent systems.

EXPERIMENT

Materials
Poly(methyl methacrylate) (PMMA, Mw=330000 Daltons), methyl methacrylate (SAFC, 98.5%), Azobisisobutyronitrile (AIBN, Fluka, 98%), acrylamide (Sigma, 99%), N,N Dimethylformamide (DMF, Sigma-Aldrich, 99%), methanol (Sigma-Aldrich, 99.7%), tetrahydrofuran (THF, Merck, 99%), dimethyl sulfoxide (DMSO, Sigma, 99%).

Wall material synthesis
PMMA as an outer wall material was synthesized by free radical polymerization of methyl methacrylate (30 ml) in presence of AIBN (1 g, initiator to monomer molar ratio of 2%) as the radical initiator in the medium of THF (50 ml) at 65°C. Polyacrylamide (PAAm) as a middle wall material was synthesized by dispersion polymerization of acrylamide monomer (30 gr) in methanol (100 ml) by using AIBN (1 g, initiator to monomer molar ratio of 1.5 %) as an initiator at 65°C.

Solutions preparation for multi-axial electrospinning
The outer layer solution was prepared by 20 wt% of PMMA in DMF. The middle wall solution contained 20 wt% PAAm in deionized water or a mixture of water and different solvents. Solutions were stirred for 24 h prior to electrospinning to obtain homogeneous solutions. Then, the prepared solutions were loaded independently into the concentric nozzles and each syringe pump controlled the flow rates. The electrospinning apparatus purchased from Yflow Company was used for electrospinning experiments.

Characterization
The surface morphologies of co-axial hollow fibers were analyzed by a Leo Supra 35VP Field Emission Scanning Electron Microscope (SEM). Elemental analysis of fibers was performed by using Energy-Dispersive X-Ray (EDX) analyzing system. The functional groups of each layer were investigated by Netzsch Fourier Transform Infrared Spectroscopy (FTIR). Thermal behaviors of fibers were examined by Netzsch Thermal Gravimetric Analyzer (TGA) by a

10°C/min scanning rate under nitrogen atmosphere. Samples were dried for 24 h at room temperature under vacuum to remove moisture and residual solvents before TGA analysis.

DISCUSSION

Solvent Selection Based on Hansen's Solubility Parameters

Miscibility and compatibility of solutions in multi-axial electrospinning process have a great effect on the final morphology of layers. Solutions with less miscibility lead to distinct wall formation whereas solutions with partial miscibility cause diffused wall morphology in the fiber structure [8]. Hansen's Solubility Parameters method is a powerful tool to control and monitor the compatibilities of solvent and solute using tabulated interactions of molecules in the form of polar (δ_p), dispersive (δ_d), and hydrogen bonding (δ_h) components [9]. Two-dimensional graphical representation of these parameters for our system is created by combining the polar (δ_p) and dispersive (δ_d) components into a new parameter of $\delta_v = (\delta_d^2 + \delta_p^2)^{1/2}$ which is plotted against δ_h. Radius of interaction (R) provides a solubility circle in two-dimensional diagram and the solvents which are placed into the circle indicate the solubility of solute. In the case of PMMA as an outer wall material, the coordinates of the center of the solubility circle are $\delta_v =$ 21.35 MPa$^{1/2}$, $\delta_h = 7.5$ MPa$^{1/2}$ and the radius of interaction is 8.6 MPa$^{1/2}$ (Figure 1a). Among the solvents located in the solubility circle of PMMA, DMF is one of suitable solvents for electrospinning process.

Water as the main solvent of PAAm is used to prepare middle wall solution. Different co-solvents with various ratios are utilized to tailor the interaction of middle and outer wall solutions (Table I). Hansen's solubility parameter of solvent mixtures is calculated by using $\delta_n^{Mix} = \sum a_i \delta_n^{i}$ equation where n represents the parameter type (p, d, or h) and a_i is the volume fraction of solvent i. The solubility map of PAAm shown in Figure 1b is plotted by using this equation. In solubility map of PAAm, green points indicate soluble regions, yellow points show turbid solution formation and red points point out immiscible solvents.

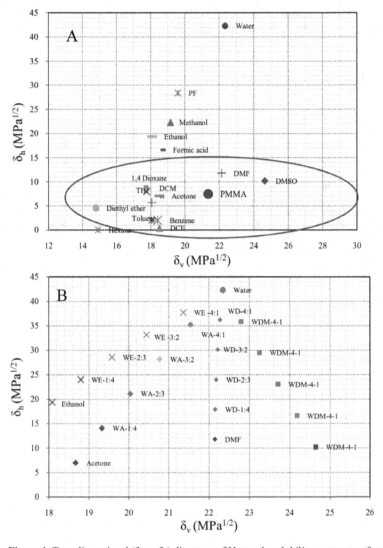

Figure 1. Two-dimensional (δ_v vs δ_h) diagrams of Hansen's solubility parameters for various solvents and solvent mixtures **(a)** PMMA solubility and **(b)** PAAm solubility

Table I. Hansen's Solubility Parameters and solubility status of PAAm in different solvent systems

Solvent system	Code	Ratio	δ_d	δ_p	δ_v	δ_h	Solubility
Water: DMF	WD-4:1	4:1	15.9	15.5	22.3	36.2	Dissolved
Water: DMF	WD-3:2	3:2	16.3	15.0	22.2	30.1	Dissolved
Water: DMF	WD-2:3	2:3	16.7	14.6	22.2	24	Dissolved
Water: DMF	WD-1:4	1:4	17.0	14.1	22.1	17.9	Turbid solution
DMF	DMF	-	17.4	13.7	22.1	11.8	Undissolved
Water: Acetone	WA-4:1	4:1	15.5	14.9	21.5	35.2	Dissolved
Water: Acetone	WA-3:2	3:2	15.5	13.7	20.8	28.2	Turbid solution
Water: Acetone	WA-2:3	2:3	15.5	12.6	20.0	21.1	Turbid solution
Water: Acetone	WA-1:4	1:4	15.5	11.5	19.3	14.1	Undissolved
Acetone	Acetone	1	15.5	10.4	18.7	7	Undissolved
Water:Ethanol	WE-4:1	4:1	15.6	16	22.3	42.3	Dissolved
Water:Ethanol	WE-3:2	3:2	15.64	14.6	21.4	37.7	Dissolved
Water:Ethanol	WE-2:3	2:3	15.6	13.1	20.4	33.1	Dissolved
Water:Ethanol	WE-1:4	1:4	15.72	11.7	19.6	28.5	Turbid solution
Ethanol	Ethanole	1	15.76	10.2	18.8	23.9	Undissolved
Water:DMSO	WDM-4:1	4:1	15.6	16	22.3	42.3	Dissolved
Water:DMSO	WDM-3:2	3:2	16.2	16.1	22.8	35.9	Dissolved
Water:DMSO	WDM-2:3	2:3	16.7	16.2	23.3	29.5	Dissolved
Water:DMSO	WDM-1:4	1:4	17.38	16.2	23.7	23.0	Dissolved
DMSO	DMSO	1	17.8	16.3	24.2	16.6	Undissolved

Surface morphology of multi-walled electrospun fibers

SEM images reveal the formation of electrospun co-axial hollow fibers produced by different pairs of solutions with different compatibilities in Figure 2. Figure 2a shows the fibers fabricated by using PAAm in water as a middle wall solution and PMMA in DMF as an outer wall solution. These solution pairs are located in longer distance of Hansen space which provides less affinity of wall solutions and results in sharp and smooth interface between middle and outer walls. In addition, PMMA, a shell of electrospun fibers, provides a brittle structure whereas PAAm shows resistance against the breakage and thus elongation of the fibers are observed. When the middle wall solution is prepared by water/DMF binary mixture, the diffusion of walls starts and the hollow structure and the boundaries of fiber disappear (Figure 2b and 2c). Therefore, an increase in the miscibility of wall solutions directly increases the interactions of walls during electrospinning.

Figure 2. SEM images of co-axial hollow electrospun fibers fabricated by different middle wall solutions **(a)** water and **(b)** and **(c)** mixture of water/DMF (volume ratio 3:2) in different regions.

Structural analysis of multi-walled electrospun nanofibers

The formation of co-axial hollow fibers is observed by monitoring the functional groups of polymeric walls. Figure 3 exhibits electrospun co-axial hollow fibers with PMMA as an outer wall and PAAm as a middle wall. For PMMA polymer, absorption bands at 2950 cm^{-1} and 1745 cm^{-1} indicate C-H and C=O stretchings, respectively[10]. For PAAm polymer, asymmetric and symmetric NH stretchings of NH_2 contributed to absorption bands at around 3300 cm^{-1} [11]. EDX results show that co-axial hollow fiber includes 56% carbon, 30% oxygen and 14% nitrogen. The nitrogen content in the fiber indicates the presence of PAAm in fiber structure.

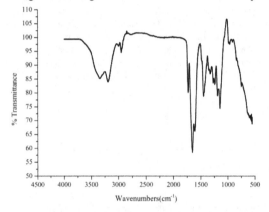

Figure 3. FTIR spectrum of co-axial hollow fiber with PMMA as an outer wall and PAAm as a middle wall

The thermal stabilities of the PMMA/PAAm co-axial hollow fibers are evaluated by means of TGA and DTA thermograms. Figure 4 exhibits TGA and DTA curves of PMMA and PAAm polymers and co-axial hollow fiber. Neat polymer radically prepared PMMA shows three steps of weight loss. At first step, PMMA loses 4% of its weight between 175-225°C due to chain scissioning of head-to-head unstable and sterically hindered linkages [12]. The second stage of degradation with weight loss of 34% is observed between 250-325°C due to scissioning of unsaturated ends (resulting from termination by disproportionation). In the last step, 62% of polymer weight is lost between 325-450°C described by random scissioning within the polymer chain [13]. In the case of neat PAAm, two stages of degradation are observed with 18% weight loss between 225-350°C because of amide side-groups decomposition and, 56% weight loss in the range of 350-500°C due to backbone decomposition [14]. The weight loss curve of co-axial hollow fibers appear between PMMA and PAAm (Figure 4a). As a result, FTIR and TGA analyses prov the successful formation of co-axial hollow fibers with different wall polymers during multi-axial electrospinning process.

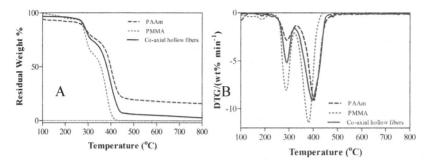

Figure 4. (a) TGA curves of PMMA, PAAm and co-axial hollow fibers and **(b)** differential thermal analyses of PMMA, PAAm and co-axial hollow fibers. Correct image

CONCLUSIONS

This study shows an optimized and quantified procedure for solvent selection and morphological control in the fabrication of multi-axial electrospun fibers by using Hansen's solubility parameters and solvent blending techniques. The distinction and diffusion of walls in the structure of fiber are tailored by changing solvent ratios in the middle wall solution. The diffusion of walls and the changes in fiber morphology are monitored by increasing the miscibility of solutions. The type of the material used in walls affect the brittleness of the fibers. PAAm, used as a middle wall in fiber structure, prevents the complete breakage of the fibers. Consequently, these co-axial hollow fibers could be applied in wide range of applications such as catalysis, hydrogen storage, self-healing, water filtration, and biomedical systems.

ACKNOWLEDGMENTS

The authors gratefully acknowledge financial support from the Scientific and Technical Research Council of Turkey (TUBITAK) Project No: 112M312/COST MP1202 and travel support from International Institute for Complex Adaptive Matter (I2CAM).

REFERENCES

[1]   E. Ozden-Yenigun, E. Simsek, Y. Z. Menceloglu, and C. Atilgan "Molecular basis for solvent dependent morphologies observed on electrosprayed surfaces," *Physical Chemistry Chemical Physics,* vol. 15, pp. 17862-17872, 2013.

[2]   M. M. Demir, M. A. Gulgun, Y. Z. Menceloglu, B. Erman, S. S. Abramchuk, E. E. Makhaeva, A. R. Khokhlov, V. G. Matveeva, and M. G. Sulman "Palladium Nanoparticles by Electrospinning from Poly(acrylonitrile-co-acrylic acid)−PdCl2 Solutions. Relations between Preparation Conditions, Particle Size, and Catalytic Activity," *Macromolecules,* vol. 37, pp. 1787–1792, 2004.

[3]   K. H. K. Chan and M. Kotaki, "Fabrication and morphology control of poly(methyl methacrylate) hollow structures via coaxial electrospinning" *Journal of Applied Polymer Science,* vol. 111, pp. 408–416, 2009.

[4]   V. Kalra, J. H. Lee, J. H. Park, M. Marquez, and Y. L. Joo "Confined Assembly of Asymmetric Block-Copolymer Nanofibers via Multiaxial Jet Electrospinning," *Small,* vol. 5, pp. 2323–2332, 2009.

[5]   H. Chen, N. Wang, J. Di, Y. Zhao, Y. Song, and L. Jiang "Nanowire-in-Microtube Structured Core/Shell Fibers via Multifluidic Coaxial Electrospinning" *Langmuir,* vol. 26, pp. 11291–11296, 2010.

[6]   W. Liu, C. Ni, D. B. Chase, and J. F. Rabolt "Preparation of Multilayer Biodegradable Nanofibers by Triaxial Electrospinning" *ACS Macro Letters,* vol. 2, pp. 466–468, 2013.

[7]   D. Han and A. J. Steckl "Triaxial Electrospun Nanofiber Membranes for Controlled Dual Release of Functional Molecules," *ACS Applied Materials & Interfaces,* vol. 5, pp. 8241–8245, 2013.

[8]   Z. Kurban, A. Lovell, S. M. Bennington, D. W. K. Jenkins, K. R. Ryan, M. O. Jones, N. T. Skipper, and W. I. F. David "A Solution Selection Model for Coaxial Electrospinning and Its Application to Nanostructured Hydrogen Storage Materials," *The Journal of Physical Chemistry C,* vol. 114, pp. 21201–21213, 2010.

[9]   C. M. Hansen, *Hansen Solubility Parameters: A User's Handbook, Second Edition* Boca Raton, FL: CRC Press, 2007.

[10]   K. Kaniappan and S. Latha "Certain Investigations on the Formulation and Characterization of Polystyrene / Poly(methyl methacrylate) Blends," *International Journal of ChemTech Research,* vol. 3, pp. 708-717, 2011.

[11]   R. Murugan, S. Mohan, and A. Bigotto "FTIR and Polarised Raman Spectra of Acrylamide and Polyacrylamide" *Journal of the Korean Physical Society,* vol. 32, pp. 505-512, 1998.

[12]   T. Kashiwagi, A. Inaba, J. E. Brown, K. Hatada, T. Kitayama, and E. Masuda "Effects of weak linkages on the thermal and oxidative degradation of poly(methyl methacrylates)," *Macromolecules,* vol. 19, pp. 2160–2168, 1986.

[13]   M. Ferriol, A. Gentilhomme, M. Cochez, N. Oget, and J. L. Mieloszynski "Thermal degradation of poly(methyl methacrylate) (PMMA): modelling of DTG and TG curves," *Polymer Degradation and Stability,* vol. 79, pp. 271–281, 2003.

[14]   A. Saeidi, A. A. Katbab, E. Vasheghani-Farahani, and F. Afshar "Formulation design, optimization, characterization and swelling behaviour of a cationic superabsorbent based on a copolymer of [3-(methacryloylamino)propyl]trimethylammonium chloride and acrylamide," *Polymer International,* vol. 53, pp. 92–100, 2004.

Mater. Res. Soc. Symp. Proc. Vol. 1621 © 2014 Materials Research Society
DOI: 10.1557/opl.2014.71

A Novel Injectable Chitosan Sponge Containing Brain Derived Neurotrophic Factor (BDNF) to Enhance Human Oligodendrocyte Progenitor Cells' (OPC) Differentiation

Mina Mekhail[1], Qiao-Ling Cui[2], Guillermina Almazan[3], Jack Antel[2], Maryam Tabrizian[1]

[1] Biomedical Engineering, Faculty of Medicine, McGill University, Montreal, QC, Canada
[2] Neurology and Neurosurgery, Faculty of Medicine, McGill University, Montreal, QC
[3] Pharmacology and Therapeutics, Faculty of Medicine, McGill University, Montreal, QC

ABSTRACT

We developed a rapidly-gelling chitosan sponge crosslinked with Guanosine 5'-Diphosphate (GDP). GDP has not been previously explored as an anionic crosslinker, and it was used in this application since the nucleoside guanosine has been shown to improve remyelination *in situ*, and thus its presence in the sponge composition was hypothesized to induce Oligodendrocyte Progenitor Cells' (OPC) differentiation. In addition to the chemical composition tailored to target OPCs, the developed chitosan sponge possesses a wide range of desirable physicochemical properties such as: rapid gelation, high porosity with interconnected pores, moduli of elasticity resembling that of soft tissue and cytocompatibility with many cell types. Moreover, protein encapsulation into the sponges was possible with high encapsulation efficiencies (e.g. BMP-7 and NT-3). In this study, BDNF was encapsulated in the chitosan sponges with an encapsulation efficiency greater than 80% and a sustained release over a 16-day period was achieved. We demonstrate here for the first time, the attachment of human fetal OPCs to the sponges and their differentiation after 12 days of culture. Overall, this newly-introduced injectable sponge is a promising therapeutic modality that can be used to enhance remyelination post-spinal cord injuries.

INTRODUCTION

Approximately 2.5 million people world-wide are living with paralysis caused by spinal cord injuries (SCI). Demyelination, or the loss of myelin sheaths surrounding axons, is one of the adverse outcomes that occurs early post-SCI due to oligodendrocyte (OG) death [1]. Demyelinated axons are not capable of transmitting action potentials, and will therefore degenerate if myelin is not re-instated. OPCs are endogenous precursors that migrate to the site of injury and remyelinate axons. However, the myelin produced is thinner compared to normal myelin, and is thus insufficient. Failure to fully remyelinate has been attributed to the inability of OPCs to fully differentiate into mature OGs [2]. The aim of this work was to address this issue through developing a rapidly-gelling, injectable chitosan sponge that is crosslinked using Guanosine 5'-diphosphate (GDP) to promote human OPC differentiation [3]. The rapid gelation ($T_{gel} < 2$ seconds) occurs due to the electrostatic interactions between cationic amine groups in chitosan and anionic phosphate groups in GDP. Guanosine (present in GDP) has been shown to induce remyelination post-SCI in animal models and was thus hypothesized to improve OPC attachment and differentiation on the sponges *in vitro*. Moreover, OPCs have been shown to be mechano-sensitive, and therefore having a biomaterial with close to native mechanical properties is highly desirable for their attachment and differentiation [4]. The compressive moduli of elasticity of the sponges have been assessed in a previous publication using a spherical indenter and a miniature loading stage [3]. The moduli of elasticity ranged from 0.87 ± 0.093 MPa and 0.43 ± 0.048 depending on the sponge formulation, which is very close to the stiffness of the spinal cord reported in the literature [5, 6]. To further promote differentiation, BDNF was encapsulated in the sponges and was released in a control manner [7]. Here we demonstrate for the first time the attachment of human fetal OPCs to the sponges and their differentiation, which was demonstrated by GalC expression. Moreover, it is important to mention that cellular preparations used were derived from 15 to 17 gestational week fetuses, and are not a pure population of OPCs, but rather only 30% of the cells are expected to be PDGFR+ (an OPC marker) [7]. Therefore, OPCs expected to differentiate will also be less than 30%.

EXPERIMENTAL DETAILS

Preparation of GDP-Crosslinked Sponges and BDNF Encapsulation

Chitosan (High Molecular Weight) at 3 and 6 mg/ml concentrations was dissolved in a 0.06M HCl solution in distilled water. The pH of the 3 mg/ml chitosan solution was raised to 5, and the pH of the 6 mg/ml solution was raised to 6 using a 1M sodium bicarbonate solution; these two solutions were given the acronyms C3PH5 and C6PH6 respectively [3]. The solutions were sterilized by filtration through 0.22 μm syringe filters and 1.7 ml from each solution was placed into Lo-Bind Eppendorfs. BDNF (0.1 μg) was then added to 1.7 ml chitosan solution and mixed thoroughly. GDP was dissolved in distilled water to a concentration of 100 mg/ml and 0.3 ml of GDP was supplemented into the chitosan solutions (final volume of 2 ml). The chitosan sponges formed in less than 2 seconds and were easily handled using forceps.

Measuring Encapsulation Efficiency and Cumulative Release Kinetics

After gelation, the sponge was removed from the Eppendorf and placed in another LoBind Eppendorf containing 0.5 ml PBS supplemented with 0.1% BSA. Release kinetics was investigated at 37°C and under mild agitation. The supernatant was removed at the designated days and stored at -20°C; fresh solution was added at each time point. In order to calculate the encapsulation efficiency, the initial solution where the sponge formed was centrifuged to pellet the sponge debris and the supernatant was removed. The concentration of un-encapsulated BDNF in the supernatant was calculated using an ELISA kit. At day 7 and 16, the samples collected to measure the release kinetics were thawed and ELISA was used to determine the BDNF concentrations. The encapsulation efficiency was calculated using Equation1, where $W_{initial}$ represents initial weight of BDNF used, and W_{free} is weight of BDNF present in the supernatant after the sponge formation. The cumulative release was calculated by the summation of percent release at each time point (W_t) as shown by Equation 2.

$$EE\ (\%) = \frac{W_{initial} - W_{free}}{W_{initial}} \times 100 \tag{1}$$

$$CR\ (\%) = \sum_{t=1}^{16} \left(\frac{W_t}{W_{initial}} \times 100 \right) \tag{2}$$

Extraction, Purification and Culturing of Human Fetal OPCs

Human fetal central nervous system tissue obtained from 15-17 gestational week embryos was provided by the Human Fetal Tissue Repository (Albert Einstein College of Medicine, Bronx, NY). PDGFα+ cells were selected with a monoclonal antibody mouse anti-human CD140a followed by rat anti mouse IgG1 magnetic beads, and the purified cells were cultured on a matrix of matrigel and poly-l lysine-coated plastic cover slips. The cultures were maintained in DMEM/F12 supplemented with N1, 0.01% bovine serum albumin, 1% penicillin-streptomycin, B27 supplement, T3, PDGF, and bFGF [7]. OPCs were cultured on top of C3PH5 and C6PH6 sponges, with and without BDNF, for a total of 12 days. The first 6 days the cells were in proliferation media (DMEM/F12 supplemented with PDGF, bFGF, T3 and B27), and the next 6 days were in differentiation media (supplemented with BDNF and IGF-1). After 12 days, OPCs were fixed and stained for O4 and GalC markers. O4+ cells are OPCs, while GalC+ cells are pre-oligodendrocytes (more differentiated).

Statistical Analysis

When comparing the encapsulation efficiencies of BDNF, a one-way ANOVA at a confidence level of 95% was used since the data was normally distributed. However, for the OPC studies, the non-parametric Kruskal Wallis statistical analysis at a confidence level of 95% was used since the data was not normally distributed.

DISCUSSION

The rapid gelation of chitosan makes this injectable sponge highly desirable for both tissue engineering and drug delivery applications. Rapid sponge formation provides two essential advantages over other injectable systems: (1) it ensures localization of the sponge after injection *in situ*, and (2) it physically entraps proteins at a high encapsulation efficiency post injection. Moreover, the porosity and interconnected pores shown by SEM images (Figure 1) allow for protein release through diffusion. The ionic crosslinking that takes place between amine and phosphate groups lead to the formation of nanometer-sized polymeric aggregates (140 nm) that fused together to form the sponge. It is important to mention that using a concentration less than 3 mg/ml of chitosan yielded nanometer polymeric aggregates that did not form a continuous porous network. Therefore, a high enough chitosan concentration is needed to provide "bridging" chitosan chains that is ionically crosslinked between polymeric aggregates and form a sponge. The guanosine molecule, which does not take part in the crosslinking process, is hypothesized to promote OPC attachment, survival and differentiation. Previous evidence demonstrated the neuro-protective and oligodendrocyte-protective effects of guanosine [1]. More interestingly, it has been shown that guanosine administered post-SCI significantly enhances the remyelination process [8]. This is the key reason GDP was explored as an anionic crosslinker in this study.

Chitosan Solution   **Chitosan Sponge**

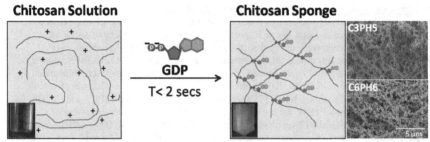

Figure 1. A schematic representation of the gelation process and a picture before and after gelation; on the right are Scanning Electron Microscopy (SEM) images of the two sponge compositions, C3PH5 and C6PH6, used in this experiment.

In order to further improve the differentiation of human OPCs, BDNF was physically entrapped in the chitosan sponge to provide controlled release to the attached OPCs in order to improve their differentiation. BDNF is a member of the neurotrophins family of proteins that have a wide range of trophic effects on both neurons and glial cells [9]. We have previously shown that BDNF in association with IGF-1 promote human fetal OPC differentiation *in vitro* [7]. The encapsulation efficiencies in both C3PH5 and C6PH6 were above 80% (Figure 2), with significantly higher encapsulation efficiency in C3PH5. The higher encapsulation efficiency was attributed to the higher water retention of C3PH5, which allows for the entrapment of more BDNF. Moreover, using a higher concentration of chitosan in C6PH6 lead to the formation of a denser chitosan network, with less water retention and thus a slower release of BDNF [3]. At day 16, C3PH5 released 19% of the encapsulated BDNF, while C6PH6 only released 11%. There is therefore a strong correlation between the chitosan concentrations and the release kinetics, since doubling the chitosan concentration almost halved the BDNF release by day 16.

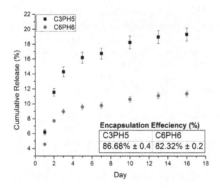

Figure 2. Cumulative release (%) of BDNF from the two sponge formulations C3PH5 and C6PH6 over a 16-day period. The encapsulation efficiencies are presented in the table, with significantly higher encapsulation efficiency in C3PH5 compared to C6PH6 ($P < 0.001$, n = 3).

It is important to mention that there is always variability between human cell preparations extracted from different fetuses. The batch used for this study had more tissue debris and less OPC differentiation compared to other preparations. Nonetheless, a comparison between human OPCs' differentiation when cultured on the sponges (+/- BDNF) and the controls is still relevant and the information is essential for designing further studies. The first observation to be made is that there were no GalC+ (O4-) oligodendrocytes on any of the sponge preparations except for C6PH6 (Figure 3A,B). This demonstrates that differentiation occurred more readily on C6PH6 sponges compared to C3PH5. The addition of BDNF to C6PH6 did not lead to any significant enhancement of differentiation. This could be attributed to the slow release of BDNF from C6PH6 (only 10% after 12 days). Moreover, the controls had 0.24% GalC+ oligodendrocytes, a very low percentage that confirms that within this batch of OPCs full differentiation into pre-oligodendrocytes was limited. The percentage of OPCs transitioning towards pre-oligodendrocytes (GalC+/O4+) was significantly higher in the controls than any of the sponge formulations. Addition of BDNF to C3PH5, however, significantly increased the percentage of GalC+/O4+ OPCs to 11% (Figure 3B), illustrating the enhanced transition towards pre-oligodendrocytes ($P < 0.01$). This could be attributed to the higher release of BDNF (18% after 12 days) from C3PH5 as compared to C6PH6. Interestingly, the controls also had a low percentage of O4+ (GalC-) OPCs (0.91%), which could be attributed to the higher rate of differentiation on the controls. On the sponges, C6PH6 containing BDNF had the highest O4+ percentage, although was not significantly different than the rest of the sponge formulations. In order to evaluate cell viability, the cell density was calculated for all the sponge formulations (**Figure 3C**). There was no statistical difference between the different formulations observed.

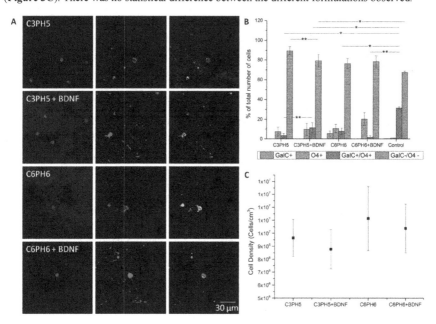

131

Figure 3. Human fetal OPC differentiation on the chitosan sponges (+/- BDNF); **A:** Confocal microscopy images illustrating GalC+/O4+ cells on the sponges (left column is DAPI, middle column is O4+, and right column is GalC+); **B:** The percentage of cells that express O4 (OPCs), GalC (pre-oligodendrocytes), GalC/O4 (transitioning OPCs), and none of the markers (n=3, $^*P<0.05$, $^{**}P<0.01$); **C:** Total cell density on the different sponge compositions.

CONCLUSIONS

This study demonstrates the ability of a novel injectable chitosan sponge to encapsulate BDNF with high encapsulation efficiency and provide a controlled release over 16 days. Human OPCs adhered to all sponge formulations and GalC+/O4+ OPCs were present on all formulations. The increased release of BDNF from C3PH5 after 12 days prompted more differentiation on those sponges as compared to C6PH6+BDNF. For future studies, incorporation of 1 µg of BDNF could be investigated for further enhancing differentiation (rather than the investigated 0.1 µg). Moreover, incorporating IGF-1 and BDNF in the sponges could produce a more pronounced enhancement of differentiation and can be more relevant for *in situ* applications. Overall, initial results from human fetal OPCs are promising, and further investigation is warranted to optimize the injectable chitosan sponges for clinical application.

ACKNOWLEDGEMENTS

The authors would like to thank the department of Biomedical Engineering for granting Mina Mekhail a BME Excellence award. Also, we would like to thank the sources of funding from CIHR and NSERC that made this work possible.

REFERENCES

[1] M. Mekhail, G. Almazan, M. Tabrizian, Progress in neurobiology, 96 (2012) 322-339.
[2] R.J. Franklin, C. Ffrench-Constant, Nat Rev Neurosci, 9 (2008) 839-855.
[3] M. Mekhail, J. Daoud, G. Almazan, M. Tabrizian, Adv Healthc Mater, (2013).
[4] A. Jagielska, A.L. Norman, G. Whyte, K.J. Vliet, J. Guck, R.J. Franklin, Stem Cells Dev, 21 (2012) 2905-2914.
[5] K. Ichihara, T. Taguchi, Y. Shimada, I. Sakuramoto, S. Kawano, S. Kawai, J Neurotrauma, 18 (2001) 361-367.
[6] R. Oakland, R. Hall, R. Wilcox, D. Barton, Proceedings of the Institution of Mechanical Engineers, Part H: Journal of Engineering in Medicine, 220 (2006) 489-492.
[7] Q.L. Cui, L. D'Abate, J. Fang, S.Y. Leong, S. Ludwin, T.E. Kennedy, J. Antel, G. Almazan, Stem Cells Dev, 21 (2012) 1831-1837.
[8] S. Jiang, M.I. Khan, Y. Lu, J. Wang, J. Buttigieg, E.S. Werstiuk, R. Ciccarelli, F. Caciagli, M.P. Rathbone, Neuroreport, 14 (2003) 2463-2467.
[9] D.M. McTigue, P.J. Horner, B.T. Stokes, F.H. Gage, J Neurosci, 18 (1998) 5354-5365.

Advances in Mechanics of Biological and Bioinspired Materials

Mater. Res. Soc. Symp. Proc. Vol. 1621 © 2014 Materials Research Society
DOI: 10.1557/opl.2014.282

DNA i-motif provides steel-like tough ends to chromosomes

Raghvendra P. Singh[1,2] , Ralf Blossey[2] and Fabrizio Cleri[1]

[1]Institut d'Electronique Microelectronique et Nanotechnologie (IEMN Cnrs - UMR 8520), University of Lille I Sciences and Technology, 59652 Villeneuve d'Ascq, France

[2]Interdisciplinary Research Institute (IRI Cnrs - USR 3078), University of Lille I Sciences and Technology, 59655 Villeneuve d'Ascq, France

ABSTRACT

We studied the structure and mechanical properties of DNA i-motif nanowires by means of molecular dynamics computer simulations. We built up to 230 nm-long nanowires, based on a repeated TC_5 sequence from NMR crystallographic data, fully relaxed and equilibrated in water. The unusual C•C+ stacked structure, formed by four ssDNA strands arranged in an intercalated tetramer, is here fully characterized both statically and dynamically. By applying stretching, compression and bending deformations with the steered molecular dynamics and umbrella sampling methods, we extract the apparent Young's and bending moduli of the nanowire, as well as estimates for the tensile strength and persistence length. According to our results, i-motif nanowires share similarities with structural proteins, as far as their tensile stiffness, but are closer to nucleic acids and flexible proteins, as far as their bending rigidity is concerned. Curiously enough, their tensile strength makes such DNA fragments tough as mild steel or a nickel alloy. Besides their yet to be clarified biological significance, i-motif nanowires may qualify as interesting candidates for nanotechnology templates, due to such outstanding mechanical properties.

INTRODUCTION

The DNA i-motif is one of the recently identified, non-standard DNA structures, which do not follow the standard Watson–Crick association rule [1-6]. In-vitro, under acidic pH conditions, the i-motif exists as a tetrameric structure, formed by four intercalated DNA strands, held together by protonated cytosine-cytosine, or C•C+, pairs. However, i–motif tetramers have also been observed *in vivo*, most notably in the terminal part of the human genes, or telomere, where rather long (50-210 bases) asymmetric G-rich and C-rich single-stranded portions of DNA are found [5,6]. Besides their possible role in the genome, still awaiting a full clarification, such DNA nanowires can be also attractive in the domain of bio-inspired materials for nanotechnologies [7-10]. Notably, various kinds of biomimetic nanowires have been already obtained from B-DNA, proteins, and even from viral particles. Electrical, optical, plasmonic features have been added to such wires by metallization, wherein metals have been "coated" or "moulded" onto the outer or inner surfaces of these biomolecular templates [11,12]. The i–motif could as well be a good candidate for nano-templating, being easily be manipulated and apparently stable over quite long time scales. However, while its molecular structure is rather well assessed, a thorough mechanical characterization of such bio-nanowire is still lacking. In a recent work [13] we calculated the main elastic moduli (Young's modulus, bending stiffness, persistence length) by means of Molecular dynamics computer simulations. In the present note, we summarize such results, and extend our investigation to the determination of the mechanical toughness of the DNA i-motif. We highlight the very peculiar character of this supramolecular structure, and speculate about the potential biological and genetic implications of this material, displaying a combination of high toughness, high strength, and good flexibility.

THEORY

Mechanical characterization of individual molecules can be performed and explored by single-molecule pulling experiments, in which molecules are stretched under the influence of a mechanical force, for example by an atomic force microscope (AFM) [14]. The detailed molecular level understanding of such pulling experiments can be greatly enhanced by the use of atomistic simulations. Steered Molecular Dynamics (SMD, [15]) is one of the best available tools, commonly used in such studies, to deduce the mechanical response of molecules subject to an external force.

In a classical AFM experiment, one end of the molecule is attached to a cantilever and the other end of the molecule is attached to a surface. Then, the cantilever is moved away from the surface with the constant velocity v, and the force response of the molecule can be measured, as being equal and opposite to the force response of the cantilever.

Similarly to what happens in a AFM experiment, in SMD simulations one end of the molecule is attached to a virtual spring, with a force constant k_c, and the other end is fixed by a harmonic restraint potential. By analogy with the AFM, the mechanical response of the molecule can be measured by moving the spring. The force response of the molecule can be calculated by the relation between spring constant and extension:

$$f = k_c (R_0 - R) \tag{1}$$

where R_0 is the total length, that is the contour length (distance between the two ends of the molecule) plus the extension of the spring, and R is the actually simulated extension of the molecule itself. The total extension is usually represented by a linear rate as:

$$R_0 = R_{fold} + vt \tag{2}$$

where R_{fold} is the initial extension in the folded state, v is the stretching velocity, and t is the time. The output of the single SMD simulation is a force-extension curve $f(R)$ of the molecule for a given stretching velocity.

In this work, SMD simulations were performed to study the elasticity under stretching and bending of the hydrated i–motif structures, using the constant–velocity or constant–force protocols of NAMD [16]. To perform SMD, four atoms (the terminal C of each strand of the i–motif tetramer) were fixed at one end of the structure. Similarly, at the opposite end, four atoms were attached to the harmonic potential, carrying the applied constant velocity. Under the applied perturbation at one end, while holding still the other end, the i–motif structure responds by developing a steady deformation, tension, compression, or bending, according to the eigenvector of the applied perturbation. The center of pull was calculated by averaging the coordinates of all the SMD atoms. The direction of pull was identified by the direction of the vector connecting the fixed and moving atoms. The center of pull of the moving atoms was attached to a dummy atom by a virtual harmonic spring. The constant velocity along the z–axis was applied to the dummy atom and the force drop between the two ends was measured. After several tests, a virtual spring constant of 1 kcal mol^{-1} Å^{-2} was adopted together with typical pulling speeds of 1 to 3 ms^{-1}, to ensure a good signal–to–noise ratio.

Bending was simulated by means of the "umbrella sampling" technique [17]. A harmonic biasing potential was used, only to simulate the action of a force at the midpoint of the nanowire (the central nucleotide tetrad, one base per chain), while holding still the two extremes. In this way we could drive the nanostructure along the bending deformation trajectory, with a pulling velocity of 5 ms^{-1} up to reaching the maximum deformation of 5 nm. The umbrella potential provided the confinement in the bent structure, while exploring the most stable conformations to extract the corresponding values of deformation energy and force along the trajectory.

DISCUSSION

Structural stability of the DNA i-motif

After a first round of classical MD simulations, we obtained the equilibrium structure of the DNA i-motif, $n[(TC_5)_4]$, constructed by the periodic repeat of the base $[(TC_5)_4]$ tetramer unit from the NMR structure [3]. Several indicators of the structural stability were assessed (Figure 1): average interplanar and intercalation base-pair distance (i.e., within each strand, and between intercalated strands, respectively); average helical twist (~12°, compared to ~34° in B-DNA); radial distribution

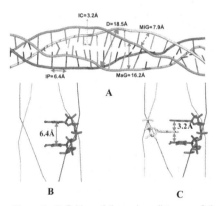

Figure 1: Definition of the various distances, of the nucleotides, in the four chains composing the i–motif tetramer. Average values obtained after the relaxation–equilibration molecular dynamics cycle described in the text. (A) The four strands show the intercalated structure of the i-motif, a tetramer composed by four independent and symmetric ssDNA strands; IP=interplanar (or base stacking) distance (see also (B)); IC=intercalation distance (see also (C)); D=diameter; MiG = minor groove width; MaG = major groove width.

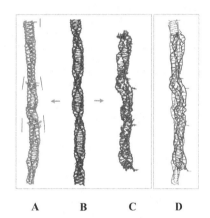

Figure 2: Comparison of the structure of the protonated vs. non-protonated i–motif tetramer after 5 ns of molecular dynamics at T=300 K. The central (B) structure in blue is the initial configuration, identical for both cases; (B) the protonated tetramer (thin bars indicate the outwards leaning Thymine bases), evolving into the correct i–motif structure; (C) the non–protonated tetramer, evolving into a disordered structure; (D), the two structures superimposed.

of water molecules and ions (Na^+ and Cl^-);sugar pucker of the ribose ring (strictly C3'-endo, compared to a mix of C2'- and C3'-endo in B-DNA); conformation of the T's, alternately leaning inwards and outwards the main cylindrical axis in agreement with experimental observation.

However, the chief indicator to demonstrate the dynamical stability of the structure was the study of the role of the resonant protonation. The comparison between the dynamical evolution of the fully protonated (n=8) 8[(TC$_5$)$_4$] i–motif and the non-protonated (np) tetramer with the same nucleotide composition, 8[(dTC$_5$)$_2$], was performed, over a time scale of 5 ns (in practical terms, the two starting structure were identical, apart from the extra proton placed between N3 nitrogen atoms of the C•C pairs.). Figure 2 shows a comparison of the structures of the i–motif (2.A) and the np–tetramer (2.C), starting from the same initial, unrelaxed 8-unit configuration (2.B). It is found that during this time, the np–tetramer crumbles, and loses the i–motif–like initial structure, becoming a rather disordered structure with a wide distribution of different interplanar distances for every base pair. By contrast, the protonated i–motif maintains a linear and straight backbone, with constant interplanar and intercalation distances, eventually adjusting to the correct, straight conformation of the well- equilibrated i-motif, and reorganizing the initial defects introduced by the artificial periodic construction. Such findings clearly support the idea of the key role of Cytosine protonation, in establishing the dynamically stable form, and structural integrity of the i–motif structure.

Mechanical properties of the DNA i-motif

We started our simulations from a fully extended 8[(TC$_5$)$_4$] i-motif nanowire, with contour length L=28.8 nm, to which the tensile deformation was applied by SMD. Figure 3 shows the typical result of a uniaxial deformation experiment for the 8[(TC$_5$)$_4$] i-motif, both in compression and in tension. The force-displacement plot under such conditions is indeed quite noisy, and displays a moderately oscillatory shape due to the non-homogeneous (also in time) molecular relaxation. However, it should be noted that the relative deformation $\Delta L/L$ over which the elastic modulus is computed is extremely small, for a typical MD simulation –0.1 to +0.3% at the slower deformation rate. In the approximation of the nanowire as a linear-elastic rigid rod of uniform density, the **Young's modulus** Y can be extracted from the linear fit of the f vs. ΔL data in Figure 3, as:

$$\frac{f}{A} = Y \left(\frac{\Delta L}{L} \right)$$
(3)

where $A=\pi R^2$ is the cross section of the nanowire (assumed cylindrical), R=0.95 nm. A clear signature of the truly linear-elastic response is the symmetry between tension and compression (positive and negative ΔL), meaning that no internal structural modifications should be induced by the deformation. The best estimate extracted from the linear fit is Y=1.8±0.5 GPa, a quite large value compared to the Young's modulus of B-DNA, which lies rather in the range of 0.35 GPa [18].

The **apparent toughness** of the *i-motif* could be probed by stretching the nanowire well beyond the linear elastic regime. Of course, with the *Amber* force field employed in our SMD simulations it is, by construction, impossible to break chemical bonds. However, an upper limit to the mechanical yielding can at least be estimated, by stretching up to the first evidences of internal structural collapse of the molecular structure. We performed extreme elongations ΔL of the 8[(TC$_5$)$_4$] *i-motif*, without detecting any signs of mechanical instability up to forces of the order of 1000 pN ($\Delta L/L \sim$ 20%) and more. Assuming a simplified definition of the principal tensile stress, averaged

138

over all the perpendicular components, and in the absence of shearing, as $<\sigma_3>=f/A$, the tensile strength σ_t (the peak value of tensile stress in the linear region of the stress-strain curve, prior to yielding) of the *i-motif* would exceed 100 MPa. Such a value of σ_t allows to estimate a **mechanical toughness**, τ, defined as the area under the stress-strain curve from 0 to σ_t:

$$\tau = \int_0^{\sigma t} \sigma_3(\varepsilon)\ d\varepsilon \qquad (4)$$

with the uniaxial strain $\varepsilon = \Delta L/L$. According to the above values, we estimate a $\tau \sim 300$ MPa, or more, notably as good as a mild steel or an aluminum alloy wire. Also in this respect, the *i–motif* appears to differ substantially from B-DNA, which is known to undergo a kind of structural, or "melting" transition above a tensile force of just ~65 pN [18], a fact that also complicates the direct experimental measurement of Y.

For a purely homogeneous cylindrical beam (or wire), the critical load to the Euler buckling instability is given as [19]:

$$F_{crit} = \frac{4\pi^2 B}{L^2} \qquad (5)$$

The mechanical meaning of the **bending stiffness** B in Eq. (5) can be appreciated by calculating the energy required to bend a wire of length L by an angle θ. If for simplicity we assume a constant curvature $\gamma = d\theta/dl$, with the radius of curvature defined as $\rho = 1/\gamma$, the bending energy (at

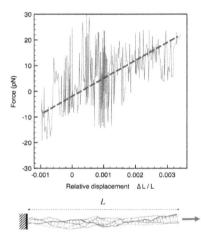

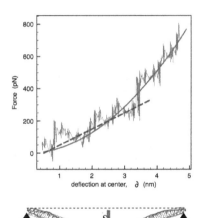

Figure 3: Force-displacement plot for uniaxial stretching deformation of the *i-motif* tetramer. Results for a constant pulling velocity of 1 ms^{-1} and a spacer spring constant of 1 kcal mol^{-1} A^{-2}. The blue dashed line is the linear fit to the average mechanical response. Below: a schematic of the stretching simulation.

Figure 4: Force-displacement plot for the bending deformation of the *i-motif* tetramer. The red continuous line is a guide to the eye (quadratic fit); the thick-dash blue line is the best linear fit at small bending. Below: a schematic of the bending simulation.

constant $\rho = L/\theta$) is [19]:

$$E_{bend} = \frac{B}{2} \int_0^L \left(\frac{d\theta}{dl}\right)^2 dl = \frac{BL}{2\rho^2} = \frac{B\theta^2}{2L} \qquad (6)$$

$l \in [0, L]$ being a variable spanning the contour length L of the wire. In free fluctuation, the thermal force $F_{th} \sim k_B T/L$ acts as a random load, with (random) compressive and bending components, whose energy is to be compared to the bending energy. For example, to bend by $\theta=30°$ a nanowire of $L=30$ nm, with a typical B value for stiffer biological polymers of $\sim 10^{-26}$ J m, the energy required according to Eq. (6) is $E_{bend} \sim 10^{-19}$ J. Since $k_B T = 4.14 \cdot 10^{-21}$ J at room temperature, it appears that only quite large thermal fluctuations, i.e. very long observation times, could lead to a reliable estimate of B by studying the fluctuation of L.

A more reliable estimate of B can be obtained from the calculation of the displacement δ at the midpoint produced by a point load f, as the derivative of the elastic energy with respect to the load (beam theory, Castigliano's theorem [19]):

$$\delta = \frac{d}{df} \left(\int_{-L/2}^{L/2} \frac{M^2(l)}{2B} dl \right) = \frac{fL^3}{48B} \qquad (7)$$

$l \in [0, L]$ being a variable spanning the contour length L of the wire, and $M(l) = \frac{1}{2} f \cdot l$ the bending moment, assumed linear along L for a straight wire. Therefore, from the fit of the linear part of the force plot as a function of δ in Figure 4, we get: $B=2.6 \pm 0.4 \cdot 10^{-26}$ N m^2.

Moreover, we can also estimate the **persistence length** from the relationship [13]:

$$\lambda_P = \frac{B}{k_B T} = 4 \pm 2 \ \mu m \qquad (8)$$

The DNA i–motif appears to belong with the class of structural polymers (such as F-actin, microtubules, keratin, etc.), as far as its Young's modulus is concerned. Moreover, it appears a nearly metallic material with its tensile strength $\sigma_t > 100$ MPa and toughness $\tau > 300$ MPa. Notably, with a value of Y in the GPa range, the i–motif is at least one order of magnitude stronger than the DNA strands by which it is made up. On the other hand, if we look at its transverse flexibility as measured by the bending stiffness B or the persistence length λ_p, the i–motif seems closer to DNA, and 2-3 orders of magnitude more flexible than F-actin or microtubules.

CONCLUSIONS

The presence of i-motif structures at the promoter regions of many oncogenes points out their potential biological significance [20-22]. Since C-rich DNA regions form i-motif structures, the opposite strand of DNA should have, in theory, the tendency to form a G-quadruplex. Such higher-order DNA structures cannot bind with proteins, nor can form active complexes for transcription, resulting in down-regulation of the close-by genes. This feature would make them a drug target of high importance.

As a relatively flexible, but very stiff and very tough nanostructure at the same time, i-motif fragments in the centromere or telomere regions of the genes could impart an unusually high toughness and rigidity to some portions of the genome, suggesting some speculations about the genetic meaning, or function, of such features. Just for the sake of argument, telomere length in white blood cells has been repeatedly observed to have an inverse correlation with blood pressure [23]; it is directly related with loss of elasticity of arterial wall [24]; and, its shortening is increasingly accepted as a predictive biomarker for cardiovascular disease [25]. It is highly speculative to ask whether such effects in leucocytes could also have a mechanical component, stemming from a higher rigidity of telomere regions, ultimately linked to the possible presence of i–motif structures. If, at least partly, chromatin integrity could be associated with the presence of long telomeres, one may wonder whether the age–related, progressive shortening of telomeres could also imply the absence of tougher i–motifs segments, someway contributing to cell aging via easier DNA degradation.

ACKNOWLEDGMENTS

Computing grants from the French Supercomputing Center IDRIS and CEA–TGCC in the frame of the PRACE 2010–030294 Project to FC and RB, are acknowledged. RPS gratefully thanks the President of the University Lille I for a collaborative, three-year PhD grant.

REFERENCES

1. J. L. Leroy, M. Gueron, J. L. Mergny and C. Helene, Nucl. Acids Res. **22** (1994) 1600
2. S. Nonin and J. L. Leroy, J. Mol. Biol. **261** (1996) 399
3. A. T. Phan and J. L. Leroy, J. Biomol. Struct. Dyn. **17** (2000) 245
4. K. S. Jin, S. R. Shin, B. Ahn, Y. Rho, S. J. Kim and M. Ree, J. Phys. Chem. B **113** (2009) 1852
5. J. Choi, S. Kim, T. Tachikawa, M. Fujitsuka and T. Majima, J. Am. Chem. Soc. **133** (2011) 16146
6. J. Smiatek, C. Chen, D. Liu and A. Heuer, J. Phys. Chem. B **115** (2011) 13788
7. Y. Wang, X. Li, X. Liu and T. Li, Chem. Commun. (Cambr) **42** (2007) 4369
8. Y. Peng, X. Wang, Y. Xiao, L. Feng, C. Zhao and J. Ren, J. Am. Chem. Soc. **131** (2009) 13813
9. X. Ren, F. He and Q. H. Xu, Chem. Asian. J. **5** (2010) 1094
10. C. Wang, Y. Du, Q. Wu, S. Xuan, J. Zhou and J. Song, Chem. Commun. (Cambr) **49** (2013) 5739
11. H. W. Fink and C. Schonenberger, Nature **398** (1999) 407
12. H. Yan, S. H. Park, G. Finkelstein, J. H. Reif and T. H. LaBean, Science **301** (2003) 1882
13. R. P. Singh, R. Blossey and F. Cleri, Biophys. J. **105** (2013)
14. M. Rief, M. Gautel, F. Oesterhelt, J. M. Fernandez and H. E. Gaub, Science **276** (1997) 1109
15. B. Isralewitz, M. Gao and K. Schulten, Curr. Opin. Struct. Biol. **11** (2001) 224
16. J. C. Phillips et al., J. Comput. Chem. **26** (2005) 1781
17. J. Kastner, J. Chem. Phys. **131** (2009) 034109
18. S. B. Smith, Y. Cui and C. Bustamante, Science **271** (1996) 795
19. J. M. Gere and S. P. Timoshenko, *Mechanics of materials,* Nelson Th., Cheltenham, UK, 1999.
20. K. Guo et al., J. Am. Chem. Soc. **129** (2007) 10220
21. N. Khan, A. Avin, R. Tauler, C. Gonzalez, R. Eritja and R. Gargallo, Biochimie **89** (2007) 1562
22. D. Sun and L. H. Hurley, J. Med. Chem. **52** (2009) 2863
23. E. Jeanclos et al., Hypertension **36** (2000) 195
24. Y. Y. Wang, H. Z. Wang, L. Y. Xie, K. X. Sui and Q.-Y. Zhang, The Aging Male **14** (2010) 27
25. J. Hoffmann and I. Spyridopoulos, Future Cardiology **7** (2011) 789

Mater. Res. Soc. Symp. Proc. Vol. 1621 © 2014 Materials Research Society
DOI: 10.1557/opl.2014.283

Scanning acoustic microscopy of biological cryosections: the effect of local thickness on apparent acoustic wave speed

Craig J. Williams[1*], Helen. K. Graham[2*], Xuegen Zhao[1], Riaz Akhtar[3], Christopher E.M. Griffiths[2], Rachel E B Watson[2], Michael J Sherratt[2] and Brian Derby[1]
[1]School of Materials, University of Manchester, Manchester, UK
[2] Institute of Inflammation and Repair, Manchester Academic and Health Sciences Centre, University of Manchester, Manchester, UK
[3] Centre for Materials and Structures, School of Engineering, University of Liverpool, UK.
*Equal contributors

ABSTRACT

Scanning acoustic microscopy (SAM), when applied to biological samples has the potential to resolve the longitudinal acoustic wave speed and hence stiffness of discrete tissue components. The heterogeneity of biological materials combined with the action of cryosectioning and rehydrating can, however, create variations in section topography. Here, we set out to determine how variations in specimen thickness influence apparent acoustic wave speed measurements

Cryosections (5μm nominal thickness) of human skin biopsies were adhered to glass slides before washing and rehydrating in water. Multiple regions (200x200 μm; n = 3) were imaged by SAM to generate acoustic wave speed maps. Subsequently co-localised 30x30 μm sub-regions were imaged by atomic force microscopy (AFM) in fluid. The images were then registered using Image J. Each pixel was allocated both a height and wave speed value before their relationship was then plotted on a scattergram. The mean section thickness measured by AFM was 3.48 ± 1.12 (SD) μm. Regional height variations influenced apparent wave speed measurements. A 3.5 μm height difference was associated with a 400 ms^{-1} increase in wave speed. In the present study we show that local variations in specimen thickness influence apparent wave speed. We also show that a true measure of wave speed can be calculated if the thickness of the specimen is known at each sampling point.

INTRODUCTION

Alterations in the gross mechanical properties of soft tissues profoundly influence tissue function. For example, the physical properties of skin are known to change with exposure to environmental factors and age [1, 2], whilst the mechanical stiffening of arterial tissue, associated with ageing and diseases such as diabetes, can lead to hypertension, stroke and heart failure [3-5]. As a consequence, in part, of the complexity of tissue structure and composition the key compositional targets of age-related mechanical remodeling remain poorly defined. Hence there is a need to develop techniques that can investigate the mechanical properties of soft tissue at micron length scales [6].

Scanning acoustic microscopy (SAM) has previously been used to characterise soft cardiovascular tissues [7] such as blood vessels [9, 10] and heart valves [11]. Compared with nanoindentation, SAM enables relatively fast image and data acquisition at a high spatial resolution (around 1 μm at 1 GHz excitation), ease of sample preparation and the ability to identify discrete tissue structures with reference to histological information [12]. Although images of tissue sections can be collected relatively easily, quantitative measurements are, however, more challenging. For example, local changes in section height (resulting from cryosectioning, rehydrating, and/or sample heterogeneity) will influence the calculated acoustic wave speed. Here we investigate if these section thickness effects can be corrected by accurately measuring the local height using AFM.

EXPERIMENTAL DETAILS

Cryosections (5 μm nominal thickness) of human skin biopsies were adhered to glass slides before washing and rehydrating in water. Multiple regions (200 x 200 μm; n = 3) were imaged by SAM using a KSI 2000 microscope (PVA TePla Analytical Systems GmbH, Herborn, Germany) modified with a custom-made data acquisition and control system [12]. The acoustic wave speed for each pixel on the image was calculated from stacks of images obtained at a range of focus distances from the surface, as previously described [12]. An assumed constant section thickness value of 5 μm was initially used to calculate the local wave speed at each pixel on the image.

Subsequently the same tissue areas were identified optically on a Catalyst AFM mounted on a Nikon fluorescence microscope (Bruker, Billerica, MA, USA). Local height data of rehydrated section (relative to the glass slide substrate) was quantified in 30x30 μm regions using PeakForce imaging and ScanAsyst Fluid tips (20nm radius of curvature, 0.7Nm^{-1} triangular cantilever). The SAM and AFM images were then registered using the ImageJ TurboReg plugin [13] allowing both height and acoustic wave speed values to be assigned to each registered pixel (Figure 1).

RESULTS AND DISCUSSION

Human Skin Cryosections

Bright field optical microscope, SAM and AFM registered images of the same tissue regions are shown in figure 1. Skin is composed of two main layers: an outer cell-rich epidermis and a supporting extracellular matrix-rich but relatively cell-poor dermis; the area where these two distinct anatomical regions abut one another is known as the dermal-epidermal junction (DEJ) and provides landmark features which facilitate image registration. In the SAM and AFM images lighter regions (brighter pixels) correspond to high acoustic wave speed and greater tissue thickness respectively. Specimen height (over all measured locations) was found to vary from 200 nm to 4 μm across the specimen whilst acoustic wave speed (again over all measured locations) varied from 1500 ms^{-1} to 2100 ms^{-1} when the specimen thickness was assumed to be constant.

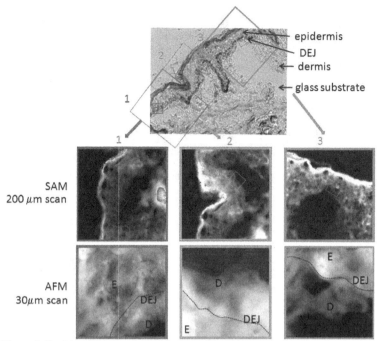

Figure 1. Registered SAM and AFM images of the skin cryosection showing the three locations used to determine the effect of local thickness changes.

Correcting acoustic wavespeed values for height (measured by AFM)

The speed of sound was related to the measured phase difference between the specimen location and a reference surface pixel, $\Delta\phi$,

$$\Delta\phi = 2\pi f \left(\frac{2d}{c_m} - \frac{2d}{c_t} \right) \tag{1}$$

where f is the acoustic frequency used for SAM imaging, d is the specimen thickness, and c_m and c_t are the wave speeds of the liquid medium and the tissue respectively. Initially the data was analysed using an assumed specimen thickness of $d^* = 5$ μm and thus we have calculated an apparent acoustic wave speed, c_t^*, assuming a constant specimen thickness. Using this equation 1 can be restated as:

$$\Delta\phi = 2\pi f\left(\frac{2d^*}{c_m} - \frac{2d^*}{c_t^*}\right) = 4\pi fd^*\left(\frac{1}{c_m} - \frac{1}{c_t^*}\right) \qquad (2)$$

The SAM measures the phase difference at a constant frequency, therefore we can combine equations 1 and 2 to correct for the assumed constant specimen thickness used in the initial analysis with

$$4\pi fd^*\left(\frac{1}{c_m} - \frac{1}{c_t^*}\right) = \Delta\phi = 4\pi fd\left(\frac{1}{c_m} - \frac{1}{c_t}\right) \qquad (3)$$

So, if the specimen thickness is known at a given location, the acoustic wave speed can be determined by:

$$\frac{1}{c_t} = \frac{1}{c_m} - \frac{d^*}{d}\left\{\frac{1}{c_m} - \frac{1}{c_t^*}\right\} \qquad (4)$$

Using equation (4) we have corrected the data for height to produce a plot of wave speed as a function of specimen thickness. Figure 2 shows the binned wave speed vs. height plots using an assumed constant section thickness (c^*) and the height corrected value (c_t) for each of the three locations indicated in Figure 1).

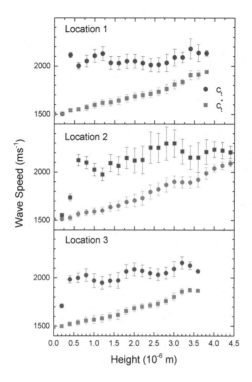

Figure 2. Acoustic wave speed vs. height plots showing the apparent wave speed at an assumed secction thicknes of 5μm and the true wave speed when corrected for AFM-derived height used,

As is clearly evident in Figure 2, the variation in wave speed across the scanned areas is found to be less when the actual thickness at each pixel is taken into account, compared with the apparent wave speed when a constant thickness was assumed. Future work will investigate other tissue types and the effect of local thickness differences.

CONCLUSIONS

Scanning acoustic microscopy can be a very powerful tool for imaging biological cryosections, however, the apparent wave speed can be influenced by changes in tissue thickness (due to the heterogeneity of tissue structure and its effect on cryosectioning) Measured values of height are required to ensure that a true measure of wave speed is recorded, when comparing local regions of a sample.

ACKNOWLEDGMENTS

The authors would like to thank the Medical Research Council for funding this work through grant reference G1001398.

REFERENCES

[1] E. C. Naylor, R. E. B. Watson, and M. J. Sherratt, "Molecular aspects of skin ageing," Maturitas, vol. 69, no. 3, pp. 249–256, 2011.

[2] M. J. Sherratt, "Tissue elasticity and the ageing elastic fibre," Age, vol. 31, pp. 305–325, Dec. 2009.

[3] P . Boutouyrie, A. I. Tropeano, R. Asmar, I. Gautier, A. Benetos, P. Lacolley, and S. Laurent, "Aortic stiffness is an independent predictor of primary coronary events in hypertensive patients:A longitudinal study," Hypertension, vol. 39, pp. 10–15, Jan. 2002.

[4] S. Aoun, J. Blacher, M. E. Safar, and J. J. Mourad, "Diabetes mellitus and renal failure: Effects on large artery stiffness," J. Hum. Hypertens., vol. 15, pp. 693–700, Oct. 2001.

[5] K. Cruickshank, L. Riste, S. G. Anderson, J. S. Wright, G. Dunn, and R. G. Gosling, "Aortic pulse-wave velocity and its relationship to mortality in diabetes and glucose intolerance:An integrated index of vascular function?" Circulation, vol. 106, pp. 2085–2090, Oct. 2002.

[6] R. Akhtar, M. J. Sherratt, J. K. Cruickshank, and B. Derby, "Characterizing the elastic properties of tissues," Mater. Today, vol. 14, no. 3, pp. 96–105, 2011.

[7] C. M. W. Daft and G. A. D. Briggs, "The elastic microstructure of various tissues," J. Acoust. Soc. Am., vol. 85, pp. 416–422, Jan. 1989.

[8] Graham HK, Akhtar R, Kridiotis C, Derby B, Kundu T, Trafford AW, Sherratt MJ "Localised micro-mechanical stiffening in the ageing aorta," Mech. Ageing Dev., vol. 132, no. 10, pp.459-67, Oct. 2011.

[9] Y. Saijo, T. Ohashi, H. Sasaki, M. Sato, C.S. Jørgensen, and S. I. Nitta, "Application of scanning acoustic microscopy for assessing stress distribution in atherosclerotic plaque," Ann. Biomed. Eng., vol. 29, pp. 1048–1053, Dec. 2001.

[10] Y. Saijo, C. S. Jorgensen, P. Mondek, V. Sefranek, and W. Paaske, "Acoustic inhomogeneity of carotid arterial plaques determined by GHz frequency range acoustic microscopy," Ultrasound Med. Biol., vol. 28, pp. 933–937, Jul. 2002.

[11] A. S. Jensen, U. Baandrup, J. M. Hasenkam, T. Kundu, and C. S. Jorgensen, "Distribution of the microelastic properties within the human anterior mitral leaflet," Ultrasound Med. Biol., vol. 32, pp. 1943–1948, Dec. 2006.

[12] X. Zhao, R. Akhtar, N. Nijenhuis, S. J. Wilkinson, L. Murphy, C. Ballestrem, M. J. Sherratt, R. E. B. Watson, B Derby, " Multi-Layer Phase Analysis: Quantifying the Elastic Properties of Soft Tissues and Live Cells with Ultra-High-Frequency Scanning Acoustic Microscopy," IEEE Trans. on Ultrason., Ferroelectr., and Freq. Control, vol. 59, no. 4, pp. 610-620, April 2012.

[13] Schneider, C.A., Rasband, W.S., Eliceiri, K.W. "NIH Image to ImageJ: 25 years of image analysis," Nature Methods, vol. 9, no. 7, pp. 671-675, Jul. 2012.

Mater. Res. Soc. Symp. Proc. Vol. 1621 © 2014 Materials Research Society
DOI: 10.1557/opl.2014.141

The Structure and the Mechanical Properties of a Newly Fabricated Cellulose-Nanofiber/Polyvinyl-Alcohol Composite

Yukako Oishi and Atsushi Hotta
Graduate School of Science and Technology, Keio University,
3-14-1 Hiyoshi, Kohoku-ku, Yokohama 223-8522, Japan

ABSTRACT

Cellulose nanofibers (Cel-F) were extracted by a simple and harmless Star Burst (SB) method, which produced aqueous cellulose-nanofiber solution just by running original cellulose beads under a high pressure of water in the synthetic SB chamber. By optimizing the SB process conditions, the cellulose nanofibers with high aspect ratios and the small diameter of ~23 nm were obtained, which was confirmed by transmission electron microscopy (TEM). From the structural analysis of the Cel-F/PVA composite by the scanning electron microscopy (SEM), it was found that the Cel-F were homogeneously dispersed in the PVA matrix. Considering the high molecular compatibility of the cellulose and PVA due to the hydrogen bonding, a good adhesive interface could be expected for the Cel-F and the PVA matrix. The influences of the morphological change in Cel-F on the mechanical properties of the composites were analysed. The Young's modulus rapidly increased from 2.2 GPa to 2.9 GPa up to 40 SB treatments (represented by the unit Pass), whereas the Young's modulus remained virtually constant above 40 Pass. Due to the uniform dispersibility of the Cel-F, the Young's modulus of the 100 Pass composite at the concentration of 5 wt% increased up to 3.2 GPa. The experimental results corresponded well with the general theory of the composites with dispersed short-fiber fillers, which clearly indicated that the potential of the cellulose nanofibers as reinforcement materials for hydrophilic polymers was sufficiently confirmed.

INTRODUCTION

Cellulose nanofibers are promising materials for the reinforcement of polymer composites, and the fabrication of cellulose nanofibers has been extensively investigated. It is, however, known to be quite challenging to extract cellulose nanofibers from original cellulose beads because of the strong interfacial interactions functioning between cellulose molecular chains due to hydrogen bondings. Therefore, nanofabrication processes of cellulose generally have multiple processing steps which occasionally required the use of toxic solvents or catalysts [1-4].

SB system developed by Sugino Machine Limited, Japan is one of the new simple methods fabricating cellulose nanofibers with harmless solvents. The system puts low stress on the environment because it only needs water jet, and the aqueous solution of cellulose nanofibers is easily produced just by running cellulose powder through the chamber with a high pressure of water [5]. Furthermore, the cellulose suspension obtained by SB method showed uniform dispersibility.

In this study, the synthesis of the composites by SB cellulose nanofibers has been first reported, and the mechanical properties and structures of synthesized Cel-F and the Cel-F/PVA composites were analyzed. The homogenous solution of Cel-F with the high aspect-ratio and the small diameter of ~23 nm, was made by increasing the SB treatment cycles (i.e. the Pass number). Furthermore, the cellulose nanofibers exhibited fine dispersibility in the PVA composites, improving the mechanical properties of the composites: the Young's modulus of the composites increased about 48% as compared with that of pure PVA. The results corresponded well with the theory of the short-fiber filler composite, which indicated that the advantage of the use of cellulose nanofibers as reinforcement fillers for hydrophilic PVA to enhance the mechanical properties of the composites was sufficiently confirmed. The cellulose/PVA composite is expected to contribute to the improvement of the mechanical properties of e.g. PVA-related biomaterials and food-packaging materials with possible cost reduction.

EXPERIMENTS

Materials

Cellulose beads were supplied from (*Nippon Paper Chemicals CO., LTD.*). The cellulose beads were mainly made of pulp being extracted from e.g. wood, plants, grasses, and straw. Polyvinyl alcohol (PVA) (Junsei Chemical Co., Ltd.) hydrolyzed at 99.0 mol% with the molecular weight of M_w=1500 was used as matrix. Purified water was supplied from Wako Pure Chemical Industries, Ltd.

Fabrication of cellulose nanofibers dispersed in water

Cellulose powder is dispersed in water and put into the SB chamber, and the liquid cellulose suspension was ejected from a nozzle at a very high pressure. The nozzle branches into a pair of nozzles to crash and pulverize polymer in the solution at a high pressure of 170 MPa and a flow rate of 600 m/sec. The number of the cycles of the SB treatment is represented by the unit "Pass". Here 10, 40, 80, and 100 Pass were selected as the SB processing numbers. The SB treatments resulted in the fabrication of nanocellulose dispersed in water.

Fabrication of Cel-F/PVA composites

PVA powder was dissolved in distilled water and stirred at 95°C for 3 hours to produce 5 wt% PVA solution with high transparency. Cel-F suspension was then mixed to compose the 5 wt% PVA solution and stirred for 1 hour. The water content of each sample was strictly controlled to be in the same condition by occasionally adding pure water. The 5 wt% Cel-F/PVA was made and the composite films of Cel-F/PVA were prepared by the solution casting method at 40°C for 24 hours. Finally, the films has been kept in a constant temperature and humidity chamber (SH-221, espec) at 25°C and at 50%RH to prevent any measurement errors among hydrophilic samples caused by the difference in the absorption of water.

Morphology of Cel-F by transmission electron microscopy (TEM)

The changes in the structures of Cel-F by treatment times from 0 Pass to 100 Pass were analyzed by transmission electron microscopy (TEM, FEI Co Ltd., TECNAI SPIRIT) at the accelerating voltage of 120 kV. For the TEM measurements, Cel-F solution was further diluted ten times by pure water. A drop of the diluted Cel-F solution was added onto a collodion film that was dried before the TEM measurements. Cel-F was observed and the widths of the fibers were estimated. The lengths of the fibers were also analyzed by TEM using sonicated Cel-F solution at 24 W for 1min to possibly unthread the fibers for the length measurements.

Dispersion of Cel-F/PVA by scanning electron microscope (SEM)

In order to observe the dispersion of Cel-F in Cel-F/PVA composite films (the Pass number ranging from 0 to 100), the scanning electron microscope (FE-SEM) (S-4700, Hitachi High-Tech) was used at a voltage of 5 kV. The fractured surface of the Cel-F/PVA composite films processed by tensile testing was examined. Osmium coating was applied for 15 sec.

Tensile strengths and dynamic mechanical properties of Cel-F/PVA measured by RSA-3

The mechanical properties of Cel-F/PVA films with different Pass numbers were analyzed. The Cel-F/PVA composite films (0, 40, and 100 Pass) were cut into a dog-bone shape (16.5 mm × 3 mm) to measure tensile strengths of the composites at the strain rate of 0.275 /min using RSA-3 (TA Instruments) at room temperature. To conduct the dynamic mechanical analysis (DMA), the RSA-3 was used with the same shape of the dog-bone specimens at the frequency ranging from 0.01 to 100 Hz.

RESULTS and DISCUSSION
Morphological changes in Cel-F

The morphology of Cel-F with different Pass numbers was investigated. Fig. 1 shows the transmission electron microscopic (TEM) images of Cel-F suspension, and the diameters of Cel-F as a function of Pass numbers. It was found from these TEM results that the diameters of Cel-F gradually became ~23 nm by increasing the Pass number and the length of the fibers reached microns. Thus the aspect-ratio became significantly high by SB method as compared with that of the original cellulose beads.

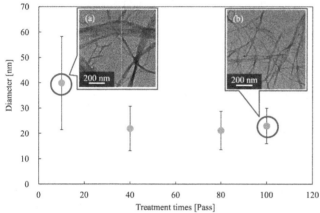

Fig. 1 TEM images of Cel-F with different Pass numbers: (a) 10 Pass and (b) 100 Pass, with the diameters of Cel-F at different Pass numbers.

The structure of Cel-F/PVA composites

Fig. 2 shows the SEM images of Cel-F/PVA composites observed for the fractured surfaces of pure PVA and the composites after tensile testing. Fig. 2(a) shows the fractured surface of the pure PVA film, and Fig. 2(b) shows the fractured surface of the 100 Pass Cel-F/PVA composite film. In Fig. 2(a), a very flat surface was observed, while in Fig. 2(b) there were many nanofibers without any voids that may degrade the mechanical properties of the composites. Strong adhesion was suggested through the SEM pictures, which may be highly due to the hydrogen bondings of both PVA and cellulose. The results, therefore, indicate the good adhesion properties between PVA and Cel-F, which may result in the enhancement of the mechanical properties of the composites.

The mechanical properties of composite

Fig. 2 shows the Young's modulus (E) of the pure PVA and the Cel-F/PVA composite films at the Cel-F concentration of 5 wt%. The E at 100 Pass showed the Young's modulus of ~ 3.2 GPa, indicating the enhancement of the Young's modulus by 48 % as compared with that of pure PVA (E_m = 2.2 GPa).

Young's modulus was considerably increased from 2.2 GPa of pure PVA to 2.9 GPa of 40 Pass composite, and it remained virtually constant after 40 Pass. This modulus behavior corresponded with the experimental data of the diameters observed by TEM (Fig.1). It was therefore considered that the diameter of Cel-F nanofibers had profound and probably decisive effects on the mechanical properties of Cel-F/PVA composite films, acting effectively as reinforcement materials.

Fig. 3 shows the tensile storage moduli (E') at 1 Hz of the pure PVA film and Cel-F/PVA composite films at 40 Pass and 100 Pass. The results of the rheological experiments indicated almost the same tendency as observed in the tensile testing. The dynamic storage modulus and the Young's modulus behaved similarly against the Pass number: with increasing the Pass number, both increased in a similar manner. Thus we may control the mechanical properties of the composites by simply changing the morphology of the Cel-F. We also analyzed these experimental mechanical results by comparing the experimental data with the theory of short-fiber filler composite based on the shear-lag model of Cox [6]. In this approach the Young's modulus E_c of the composites was given by:

$$E_c = E_f V_f \left[1 - \tanh\left(\frac{\beta l}{2}\right) \cdot \left(\frac{\beta l}{2}\right)^{-1} \right] C_a + E_m \left(1 - V_f\right)$$

where V_f is the area fraction of the fibers, E_f (=145 GPa) [4] is the Young's modulus of the fibers, E_m (= 2.2 GPa) is the Young's modulus of the matrix, C_a (= 1/6) is the coefficient of orientation for three random dimensions, l is the length of the fibers, and β is given by:

$$\beta = \frac{1}{r} \sqrt{\frac{2G_m}{E_m\left(R \cdot r^{-1} - 1\right)}}$$

where G_m is the shear modulus of the matrix, r is the radius of the fibrils, and $(R \cdot r^{-1})^2 = V_f^{-1}$. In Fig. 2, the calculated theoretical values were also plotted, which corresponded well with the experimental Young's modulus that was increased by increasing the Pass number. The increase in the Young's modulus was mainly caused by the decrease in the width of Cel-F. In fact, it was theoretically known that the radius of the reinforcement fillers along with the morphology of the composites had profound effects on the mechanical properties of the resulting composites. In the Cel-F/PVA composites, since the length of Cel-F was almost constant regardless of the Pass number, the enhancement of the aspect ratio of Cel-F was primarily due to the decrease in the radius of Cel-F, hence the decrease in the width of Cel-F. The Young's modulus at 100 Pass was a little bit even higher that the theoretical value, which may be due to the fact that Cel-F was not only homogeneously dispersed but also well entangled with each other to even raise the modulus. To summarize, SB treatment is a promising method to produce cellulose nanofibers with high aspect ratios, and the well-entangled nanofibers can be homogeneously dispersed in hydrophilic polymers such as PVA afterward by solution casting.

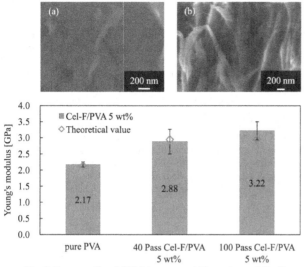

Fig. 2 Cross-sectional SEM images and Young's moduli of pure PVA and Cel-F/PVA composites at the concentration of 5 wt%: (a) pure PVA and (b) 100 Pass Cel-F/PVA

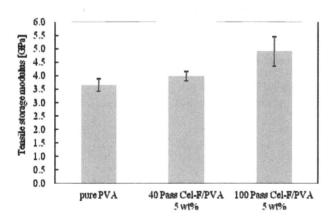

Fig. 3 Tensile storage moduli of pure PVA and Cel-F/PVA composites at the concentration of 5 wt% against the Pass number.

CONCLUSIONS

The cellulose nanofiber and its PVA composite were analyzed in terms of the structures and the mechanical properties. A high aspect-ratio cellulose nanofiber (Cel-F) with the diameter of ~23 nm was synthesized by a newly developed water-jet nano-fabrication process, called SB method. The diameter of Cel-F was efficiently decreased to ~23 nm with the increase in the Pass number. From the structural analyses of Cel-F, the detailed fabrication process of the fibers was revealed: the fibrils of cellulose nanofibers were split off from cellulose powder and the fibrils were further torn off to become cellulose nanofibers down to the diameter of ~23 nm. According to the studies of Cel-F/PVA composites, it was found by the SEM micrographs that Cel-F were homogeneously dispersed in PVA matrix, showing good adhesion to PVA. Due to the uniform and homogeneous dispersibility of the Cel-F in PVA, the Young's modulus and the storage modulus of the 100 Pass Cel-F/PVA composite at the concentration of 5 wt% increased by 48% as compared with that of pure PVA. The experimental results of Young's modulus agreed well with the theory of short-fiber filler composites, suggesting the potential utilization of the cellulose nanofibers as reinforcement materials for the enhancement of the mechanical properties of polymers.

REFERENCES

1. Kamphunthong W, Hornsby P, and Sirisinha K., Journal of Applied Polymer Science, **125** (2), 1642-1651 (2012).
2. Tang CY and Liu HQ., Composites Part a-Applied Science and Manufacturing, **39** (10), 1638-1643 (2008).
3. Tang CY, Wu MY, Wu YQ, and Liu HQ., Composites Part a-Applied Science and Manufacturing, **42** (9), 1100-1109 (2011).
4. Fujisawa S, Ikeuchi T, Takeuchi M, Saito T, and Isogai A., Biomacromolecules, **13** (7), 2188-2194 (2012).
5. Watanabe Y, Kitamura S, Kawasaki K, Kato T, Uegaki K, Ogura K, and Ishikawa K., Biopolymers, **95** (12), 833-839 (2011).
6. Cox HL., British Journal of Applied Physics, **3** (MAR), 72-79 (1952).

Mater. Res. Soc. Symp. Proc. Vol. 1621 © 2014 Materials Research Society
DOI: 10.1557/opl.2014.368

Modular Peptide-Based Hybrid Nanoprobes for Bio-Imaging and Bio-Sensing

Banu Taktak Karaca,[1,2] James Meyer,[1] Sarah VanOosten,[1] Mark Richter,[3] Candan Tamerler[1,4]

[1]Bioengineering Research Center (BERC) & Bioengineering Program, University of Kansas, Lawrence, KS 66045, USA
[2]Department of Molecular Biology and Genetics, Istanbul Technical University, Istanbul 34469, Turkey
[3]Department of Molecular Biosciences, University of Kansas, Lawrence, KS 66045, USA
[4]Department of Mechanical Engineering, University of Kansas, Lawrence, KS 66045, USA

ABSTRACT

The self-organization of functional proteins directly onto solid materials is attractive to a wide range of biomaterials and systems that need to accommodate a biological recognition element. In such systems, inorganic binding peptides may be an essential component due to their high affinity and selective binding features onto different types of solid surfaces. This study demonstrates a peptide-enabled self-assembly technique for designing well-defined protein arrays over a metal surface. To illustrate this concept, we designed a fusion protein that simultaneously displays a red fluorescence protein (DsRed-monomer), which is highly selective for copper ions, and a gold binding peptide AuBP. The peptide tag, AuBP, self-directs the organization of DsRed-monomer protein onto a gold surface and forms arrays built upon an efficient control of the organic/inorganic interface at the molecular level. The peptide-assisted design offers a modular approach for fabrication of fluorescent-based protein arrays with copper ion sensing ability.

INTRODUCTION

The diverse properties of natural materials including exceptional mechanical properties have always been at the core of bioinspired materials [1-4]. However, there have been challenges in translating the design strategies used by Nature to the engineering of new materials. The way Nature evolves the biological systems serves to bring multiple distinguishing functions that are built upon bio-molecular machinery. One of the unique features in this approach for engineers is probably their inherent ability to self-assemble over multiple length scales. Among different biomolecules, proteins play an important role in bio-molecular systems by performing diverse functions built into them due to their functional specificity and their precise molecular recognition properties [3-10]. In the last decade, there has been a growing interest in mimicking molecular scale interactions of biomolecules at the solid material interfaces to design novel materials and systems. Particularly, biological functionalization of surfaces is increasingly attracting interest in a wide range of medical and technological applications [5-10].

Bio-integrated systems require the organization of biomolecules onto surfaces with resolutions ranging from micro to nanometer scale [10-12]. The wide variety of protein molecules and their functionalities appeals as enablers for the construction of complex and miniaturized devices such as protein arrays, biosensors and biochips [5, 6, 13-16]. Currently, there are different techniques to create protein arrays on surfaces, including the use of self-assembled monolayers (SAM) and patterning using lithography techniques [17-18]. The commonly used linkers are silanes and thiols, which bind at one end to the inorganic materials surface via silanol or thiol bonds and at the other end, have a pendant group for protein binding [18-20]. Despite their wide usage, these linkers have certain disadvantages including instability, lack of material selectivity or multilayer coverage. As an alternative to current chemical coupling protocols, the inorganic binding peptides are attracting high interest as surface specific bio-linkers [1, 3, 10-12, 21]. The inorganic binding peptides are selected from combinatorial peptide phage or cell surface display libraries by a variety of groups [3, 21-23]. This peptide-mediated surface functionalization provides the control over the assembly process in a self-directed manner. Their dissociation constant value varies within the μM to nM range while exhibiting selective material binding ability over a wide range of surfaces [3, 23]. The properties of these peptides were shown in surface functionalization, self-assembly and formation of various nanostructures [3, 7-11, 21-23]. They were also demonstrated to be efficient surface linkers as fusion partners to other biomolecules such as proteins [12, 14-16, 24-25].

Here, we engineered a multi-functional fusion protein, which consists of red fluorescence protein (DsRed), and an Au binding peptide (AuBP). In our previous studies, we reported screening of gold binding peptides (AuBPs) and confirmed their high affinity and selective binding characteristics onto the gold surfaces [11]. The AuBP2c peptide tag was utilized to enable the self-immobilization of DsRed protein onto a gold surface in a single step using micro contact printing (μCP). The red emission of DsRed protein offers new opportunities for multicolor labeling and FRET applications at the nano- to micro scale material surfaces [26]. Consequently, the engineered DsRed construct provides an easy visualization of the self-immobilized proteins on the gold surface. Additionally, DsRed was shown to have selective and reversible binding to copper ion by Rahimi et.al [27]. The copper ion plays important roles such as an enzyme co-factor, in exhibiting high ionic conductivity and being a major component in heavy metal pollution. Due to its importance in several diverse areas, several biosensors were proposed to monitor copper in recent years. Here, we investigated if we can tailor the fluorescence activity of DsRed while it is genetically conjugated to AuBP. Our results show the degree of quenching with respect to the amount of copper ion. We also demonstrated that the DsRed-AuBP protein self-organization is functional on metallic surfaces and thus of benefit to fabricating proteins arrays.

EXPERIMENTAL DETAILS

The plasmid pDsRed-Monomer (Clontech, Mountain View, CA) was used as a template for polymerase chain reaction (PCR). The primers are designed to encode the DNA sequence of the AuBP2c peptide. Completing three consecutive PCR reactions, the DNA sequence for the peptide, with a 4 amino acid linker (SGGG) and restriction enzyme sites, were inserted at the C-terminus of the DsRed-monomer sequence. The gene of DsRed-AuBP2c was inserted into *Hind* III and *Eco*R I sites of the expression vector pMALc-4X (New England Biolabs, NEB, Beverly, MA) containing the maltose binding protein (MBP) encoding gene (Figure 1a-b). The plasmid pMAL-DsRed-AuBP2c was transformed into the *Escherichia coli* 2507 cells. The cells containing the plasmid (pMAL-DsRed-AuBP2c), which encodes the DsRed-AuBP2c fusion protein, were grown in LB medium containing ampicillin at 250 rpm at 37°C up to an OD_{600} of 0.6. Protein expression was then induced using IPTG (0.3 mM) and the cells were allowed to grow for an additional 48 h at 30°C. The expressed protein was purified on an amylose resin

(NEB) column. After purification, the MBP tag was digested with Factor Xa proteolytic enzyme and removed from the DsRed-AuBP2c fusion protein using a HiTrap (GE Healthcare Bio-Sciences, Pittsburg, PA) metal chelating column. The purity was determined by SDS-PAGE using Instant Blue solution (Expedeon Inc., San Diego, CA). Protein concentrations were determined using the Bradford assay.

Different concentrations of copper were added to 1 µM of DsRed-AuBP2c protein solution. The fluorescence readings were measured using fluorescence reader by excitation of the samples at 556 nm and the resulting emission at 590 nm was recorded. DsRed protein was used as the control protein to compare the fluorescence readings with DsRed-AuBP2c.

The patterning of the proteins onto gold surfaces was performed by µCP. DsRed-AuBP2c and DsRed (50 µM) proteins were applied onto the clean polydimethylsiloxane (PDMS) stamp for 10 min. Excess protein was then removed by serial washing and the stamps were gently dried with nitrogen. The clean gold surfaces were placed in contact with the protein-loaded stamps, allowed to sit for 10 min, washed with DI water and dried with nitrogen. The DsRed-AuBP2c fusion protein pattern was imaged using fluorescence microscope with DsRed filter.

RESULTS AND DISCUSSION

Multi-functional fusion protein: DsRed-AuBP2c

We engineered a fusion protein containing two functional domains; red fluorescent protein (DsRed), and highly specific AuBP2c peptide tag. The pMALc-4X expression vector encoding the maltose binding protein (MBP) was chosen for ease of cloning, over-expression and purification of the fusion protein (Figure 1a).

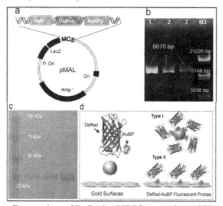

Figure 1: Vector Design, Expression of DsRed-AUBP2c protein and Hybrid NanoProbe Design. Schematic of vector construction (a), Vector constructs on agarose gel (b), line (1-3): pMal-DsRed-AuBP2c vector, line (M3): Lambda DNA/EcoRI+HindIII Marker (Fermentas), SDS-PAGE result of DsRed-AuBP2c (~32 kDa) following the affinity column purification (c), Schematic of fluorescent hybrid nanoprobes (d). DsRed protein structure (DOI:10.2210/pdb2vad/pdb) was visualized by PyMOL Software (Schrödinger, LLC).

The MBP provided a useful, monovalent scaffold for peptide-based multifunctional protein production for several reasons. First, fused proteins are purified by one-step affinity chromatography on cross-linked amylose and bound fusion proteins can be eluted using maltose (10 mM) in physiological buffer. Second, MBP has no cysteine that could form disulfide bonds with the fused peptide. Finally, there is less concern regarding potential protein solubility issues due to high solubility of MBP when used as a fusion partner. In some cases, the fusion to MBP can also promote the proper folding of the attached protein into its biologically active conformation. Additionally, the pMAL-4 vectors contain the sequence coding for the recognition site of a specific protease, located just 5′ to the polylinker insertion sites. This allows MBP to be cleaved from the protein of interest after purification.

Fusion protein-copper ion interactions

To investigate if the fluorescence intensity will be affected by the addition of copper ion, the fusion protein, DsRed-AuBP2c, was titrated with 1-200 µM copper II (Cu^{2+}) ion. The change in emission intensity at 590 nm (red) was obtained by increasing the copper concentrations. The red emission decreased by 50-90% at copper concentrations above 10 µM. The decreasing fluorescence intensity of the protein was shown in Figure 2a following the addition of the copper ion. We also followed the consequent increase in quenching against total copper concentration (Figure 2b). Binding affinity of the fusion protein for copper was quantified by fitting the conventional model provided in Equation 1 and 2 to the fluorescence data.

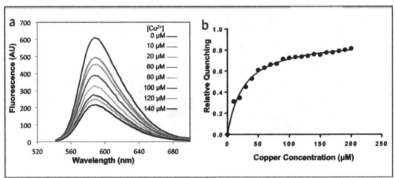

Figure 2 Titration of DsRed-AuBP2c with Cu^{2+}. Emission spectra of DsRed-AuBP2c (a) was obtained by excitation at 556 nm in the presence of 0, 10, 20, 60, 80, 100, 120, 140 µM Cu^{2+}. The change in quenching was integrated over a wavelength range of 560-650 nm and is plotted against copper concentration. 0.8 on the ordinate indicates 100% change in observed fluorescence.

$$F = F_0 + \Delta F \text{ [bound] / [P]} \quad (1)$$

where F is the measured fluoresce after normalization, F_0 is the fluorescence in the absence of copper, and ΔF is the fluorescence change caused by copper binding. The concentration of the protein-copper, [bound], is shown in Equation 2,

$$\text{[bound]} = \{ K_d + [Cu]_{tot} + [P] - \text{sqrt} ([K_d + [Cu]_{tot} - [P])^2 - 4[Cu]_{tot} [P]]\}/2 \quad (2)$$

where K_d is the dissociation constant of the copper binding sites, total copper concentration is $[Cu]_{tot}$, and $[P]$ is protein concentration. The fitted values of K_d were calculated to be 17.94 ± 1.63 µM for DsRed-AuBP2c.

Fabrication of protein array through micro-contact printing

Due to its parallel nature, micro-contact printing (µCP) has proven to be a convenient technique to pattern a variety of biological molecules with sub-micron features over a large area (> 1 cm^2) without the need for expensive lithographic equipment. In our previous work, we demonstrated that the spatial conformation of the adsorbed peptide may play an important role in its binding with surfaces. Here, we investigated the fusion protein self-organization into a well-defined array with minimum non-specific binding onto the undesired regions of the array. As shown in Figure 3, fluorescence microscopy images of protein patterned on gold substrate demonstrates the efficient organization of DsRed-AuBP2c fusion protein on gold surface resulting in an array formation compared to DsRed-monomer alone. By using the micro-contact printing (µCP) technology, we demonstrated a bio-friendly fabrication of protein arrays on a gold surface in a single step. The AuBP2c peptide tag self-directs the immobilization of fusion protein onto the gold surface, consequently the engineered fluorescence protein self-assembles onto the gold surface forming a well-defined pattern with a high patterning efficiency.

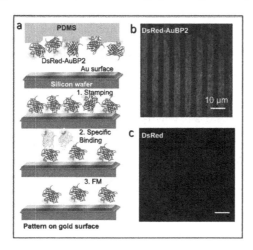

Figure 3. Protein Patterning onto a Flat Gold Surface with the Micro-Contact Printing Technique: schematics of micro-contact printing (a), fluorescent images of 50 µM DsRed-AuBP2c protein (b) and wild-type DsRed-monomer protein as control (c).

159

CONCLUSIONS

Addressable attachment of biomolecules through self-assembly processes onto solid surfaces remains a primary challenge in building biomedical materials, proteomic arrays, chips and sensors. Currently, the assembly of biomolecules onto solid substrates is accomplished using covalently bound chemical linkers including thiols and silanes. Although successfully used in many applications, their inherent limitations continue to exist due to restricted stability, lack of material specificity and lack of control over the biomolecule orientation bound to the surface. Combinatorially selected inorganic-binding peptides prove to be an efficient alternative in biological surface functionalization. These peptides not only have the molecular recognition of inorganic surfaces with high affinity but also they can be part of a functional protein using genetically engineering approaches to incorporate self-organization ability over a wide range of solid materials. Here, the results demonstrate the utility of a gold binding peptide as a molecular enabler for self-organization of a red fluorescence protein to form arrays on planar gold surfaces. We also demonstrated the copper ion monitoring ability of the engineered fusion protein following the decrease in fluorescence intensity and the reversibility of the quenching in the presence of a chelator. Overall, bio-functionalization of surfaces using engineered peptide based systems provides a viable alternative to the conventional chemical coupling to produce various protein assemblies resulting in controllable nanoscale organization.

ACKNOWLEDGMENTS

This work was supported by KU-Mechanical Engineering Funds at Bioengineering Research Center (BERC) and The Scientific and Technological Research Council of Turkey (TUBITAK) through a 2214-A Scholar Program provided for Banu Taktak Karaca.

REFERENCES

[1] Sarikaya M, Tamerler C, Schwartz DT, Baneyx FO. Materials assembly and formation using engineered polypeptides. Annual Review of Materials Research. 2004;34:373-408.
[2] Meyers MA, Chen PY, Lin AYM, Seki Y. Biological materials: Structure and mechanical properties. Prog Mater Sci. 2008;53:1-206.
[3] Tamerler C, Sarikaya M. Molecular biomimetics: Genetic synthesis, assembly, and formation of materials using peptides. Mrs Bull. 2008;33:504-10.
[4] Fratzl P, Weinkamer R. Nature's hierarchical materials. Prog Mater Sci. 2007;52:1263-334.
[5] Willner I. Biomaterials for sensors, fuel cells, and circuitry. Science. 2002;298:2407-8.
[6] Willner I, Willner B. Biomolecule-Based Nanomaterials and Nanostructures. Nano Lett. 2010;10:3805-15.
[7] Naik RR, Stringer SJ, Agarwal G, Jones SE, Stone MO. Biomimetic synthesis and patterning of silver nanoparticles. Nat Mater. 2002;1:169-72.
[8] Slocik JM, Naik RR. Biologically programmed synthesis of bimetallic nanostructures. Adv Mater. 2006;18:1988.
[9] Slocik JM, Stone MO, Naik RR. Synthesis of gold nanoparticles using multifunctional peptides. Small. 2005;1:1048-52.
[10] Briggs BD, Knecht MR. Nanotechnology Meets Biology: Peptide-based Methods for the Fabrication of Functional Materials. Journal of Physical Chemistry Letters. 2012;3:405-18.
[11] Tamerler C, Duman M, Oren EE, Gungormus M, Xiong XR, Kacar T, Parviz BA, Sarikaya M. Materials specificity and directed assembly of a gold-binding peptide. Small. 2006;2:1372-8.

[12] Hnilova M, Karaca BT, Park J, Jia C, Wilson BR, Sarikaya M, Tamerler C. Fabrication of hierarchical hybrid structures using bio-enabled layer-by-layer self-assembly. Biotechnology and Bioengineering. 2012;109:1120-30.

[13] Woodbury RG, Wendin C, Clendenning J, Melendez J, Elkind J, Bartholomew D, Brown S, Furlong CE. Construction of biosensors using a gold-binding polypeptide and a miniature integrated surface plasmon resonance sensor. Biosens Bioelectron. 1998;13:1117-26.

[14] Coyle BL, Rolandi M, Baneyx F. Carbon-Binding Designer Proteins that Discriminate between sp(2)- and sp(3)-Hybridized Carbon Surfaces. Langmuir. 2013;29:4839-46.

[15] Kacar T, Zin MT, So C, Wilson B, Ma H, Gul-Karaguler N, Jen AKY, Sarikaya M, Tamerler C. Directed Self-Immobilization of Alkaline Phosphatase on Micro-Patterned Substrates Via Genetically Fused Metal-Binding Peptide. Biotechnol Bioeng. 2009;103:696-705.

[16] Hnilova M, Khatayevich D, Carlson A, Oren EE, Gresswell C, Zheng S, Ohuchi F, Sarikaya M, Tamerler C. Single-step fabrication of patterned gold film array by an engineered multi-functional peptide. J Colloid Interf Sci. 2012;365:97-102.

[17] Haussmann A, Milde P, Erler C, Eng LM. Ferroelectric Lithography: Bottom-up Assembly and Electrical Performance of a Single Metallic Nanowire. Nano Lett. 2009;9:763-8.

[18] Fenter P, Eisenberger P, Li J, Camillone N, Bernasek S, Scoles G, Ramanarayaanan TA, Liang KS. Structure of $CH_3(CH_2)17SH$ Self-Assembled on the Ag(111) Surface: An Incommensurate Monolayer. Langmuir. 1991;7:2013-6.

[19] Leong K, Chen YC, Masiello DJ, Zin MT, Hnilova M, Ma H, Tamerler C, Sarikaya M, Ginger DS, Jen AKY. Cooperative Near-Field Surface Plasmon Enhanced Quantum Dot Nanoarrays. Advanced Functional Materials. 2010;20:2675-82.

[20] Porter MD, Bright TB, Allara DL, Chidsey CED. Spontaneously Organized Molecular Assemblies .4. Structural Characterization of Normal -Alkyl, Thiol Monolayers on Gold By Optical Ellipsometry Infrared-Spectroscopy and Electrochemistry. J Am Chem Soc. 1987;109:3559-68.

[21] Shiba K. Exploitation of peptide motif sequences and their use in nanobiotechnology. Current Opinion in Biotechnology. 2010;21:412-25.

[22] Puddu V, Perry CC. Peptide Adsorption on Silica Nanoparticles: Evidence of Hydrophobic Interactions. Acs Nano. 2012;6:6356-63.

[23] Evans JS, Samudrala R, Walsh TR, Oren EE, Tamerler C. Molecular design of inorganic-binding polypeptides. Mrs Bull. 2008;33:514-8.

[24] Hnilova M, Liu X, Yuca E, Jia C, Wilson B, Karatas AY, Gresswell C, Ohuchi F, Kitamura K, Tamerler C. Multifunctional Protein-Enabled Patterning on Arrayed Ferroelectric Materials. Acs Appl Mater Inter. 2012;4:1865-71.

[25] Yuca E, Karatas AY, Seker UOS, Gungormus M, Dinler-Doganay G, Sarikaya M, Tamerler C. In Vitro Labeling of Hydroxyapatite Minerals by an Engineered Protein. Biotechnol Bioeng. 2011;108:1021-30.

[26] Mizuno H SA, Eli P, Hama H, Miyawaki A. Red fluorescent protein from Discosoma as a fusion tag and a partner for fluorescence resonance energy transfer. Biochemistry. 2001 Feb 27;40:2502-10.

[27] Rahimi Y, Goulding A, Shrestha S, Mirpuri S, Deo SK. Mechanism of copper induced fluorescence quenching of red fluorescent protein, DsRed. Biochemical and Biophysical Research Communications. 2008;370:57-61.

Mater. Res. Soc. Symp. Proc. Vol. 1621 © 2014 Materials Research Society
DOI: 10.1557/opl.2014.284

Microstructure, Spectroscopic Studies and Nanomechanical Properties of Human Cortical Bone with Osteogenesis Imperfecta

Chunju Gu[1], Dinesh R. Katti[1], Kalpana S. Katti[1]
[1]Civil and Environmental Engineering, North Dakota State University, Fargo, ND 58102, U.S.A.

ABSTRACT

Bone is a natural protein (collagen)-mineral (hydroxyapatite) nanocomposite with hierarchically organized structure. Our previous work has demonstrated orientational differences in stoichiometry of hydroxyapatite resulting from orientationally dependent collagen-mineral interactions in bone. The nature of these interactions has been investigated both through molecular dynamics simulations as well as nanomechanical and infrared spectroscopic experiments. In this study, we report experimental studies on human cortical bone with osteogenesis imperfecta (OI), a disease characterized by fragility of bones and other tissues rich in type I collagen. About 90% of OI cases result from causative variant in one of the two structural genes (COL1A1 or COL1A2) for type I procollagens. OI provides an interesting platform for investigating how alterations of collagen at the molecular level cause changes in structure and mechanics of bone. Fourier transform spectroscopy, electron microscopy (SEM), and nanomechanical experiments describe the structural and molecular differences in bone ultrastructure due to presence of diseases. Photoacoustic-Fourier transform infrared spectroscopy (PA-FTIR) experiments have been conducted to investigate the orientational differences in molecular structure of OI bone, which is also compared with that of healthy human cortical bone. Further, in situ SEM static nanomechanical testing is conducted in the transverse and longitudinal directions in the OI bone. Microstructural defects and abnormities of OI bone were ascertained using scanning electron microscopies. These results provide an insight into molecular basis of deformation and mechanical behavior of healthy human bone and OI bone.

INTRODUCTION

Osteogenesis imperfecta (OI) is a heritable disease that is characterized by fragility of bones and other tissues rich in type I collagen. The two structural genes, COL1A1 and COL1A2, which encode the proα1(I) and proα2(I) chains of type I procollagen, are known to exhibit over 2000 distinct mutations in the event of osteogenesis imperfecta[1]. These mutations range in complexity from simple deletions, insertions, and single base substitutions that convert a codon for glycine to a codon for a bulkier amino acid. These mutations not only affect collagen molecules and other bone cell matrix components, but also affect the mineral phase with higher average mineralization density[2], smaller, less well aligned mineral crystal size, and highly packed and disoriented mineral crystals[3-5].

Our previous work has demonstrated orientational differences in stoichiometry of hydroxyapatite in healthy bone which is influenced by collagen-mineral interaction[6]. In addition extensive multiscale modeling experiments in our group[7] have also demonstrated the directional dependence of mechanics of collagen[8], the influence of the interaction[9], the role of helical structure of collagen molecules, as well as the discovery of 3rd-tiered hierarchy on the mechanics of collagen[7, 8, 10]. In the present study, we select a piece OI cortical bone (suspected to be type I OI) and study the ultrastructure, molecular differences, and mineral crystal structure as well as nanomechanical properties with the techniques of field emission scanning electron microscopy (FE-SEM), photoacoustic-Fourier transform infrared spectroscopy (PA-FTIR), and in situ FE-SEM nanoindentation. These results were also compared with those of healthy cortical bone.

EXPERIMENT

The human OI tibia was obtained from National Disease Research Interchange, PA (No apparent metabolic bone disease record, 22 years old, female) and stored in a freezer at -70°C. The OI type of this sample was not identified; yet suspected to be type I since the patient had height of 67 inches, and weight of 180lb. Specimen with thickness of about 1 mm in both transverse and longitudinal directions in the anterior area were obtained from the mid-diaphysis of the femur (Fig.1). These specimen were used for PA-FTIR characterization. Sample preparation for the SEM imaging incorporated several additional steps described elsewhere[6]. In order to perform nanoindentation, the specimen were ground and polished.

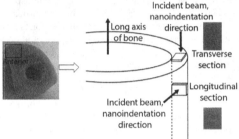

Figure 1. An OI bone sample is trimmed down to two pieces as transverse section and longitudinal section.

Microstructure of the bone specimen was studied using a JEOL JSM-7600F analytical high resolution field-emission scanning electron microscope (FE-SEM). PA-FTIR spectra were collected using a Thermo Electron, Nexus 870 spectrometer which was equipped with MTEC Model 300 photoacoustic accessory. All spectra were collected in the range of 4000-400 cm[-1], with a spectral resolution of 4 cm[-1]. In situ FE-SEM nanoindentation tests were performed using a Hysitron PI-85 nanomechanical instrument which was integrated in the JEOL JSM-7600F SEM. Nanoindentation tests were performed with static load control; 50 μN was applied as the load for the sample on the transverse section plane. An interlamellar cement band was easily observed that separates the osteon and interstitial lamellae. Thirty indents were performed inside the osteon, and ten indents outside the osteon. For the sample on the longitudinal plane (Fig.1), 15 μN, 50 μN, and 100 μN were applied with thirty indents for each load. Some surface defects were observed (e.g., pitch, cracks, pop-outs) and excluded from the nanoindentation.

DISCUSSION

The classic non-deforming OI cases either result from mutations in one COL1A1 allele (frameshift, nonsense and splice-site alteration) or substitutions for glycine by small amino acid (cysteine, alanine and serine). As the mutations are heterozygous, some of the gene products should be normal. Even in severe OI there are normal lamellar bone structure composed of normally mineralized fibrils[11]. Therefore, OI has more local structural variations within individual bones.

Microstructure of OI human cortical bone

Healthy bone has a comparatively homogenous, well attached mineralized fibrils structure[11]. However, the surfaces of OI bone specimen fractured in liquid N_2 show significantly different FE-SEM images than the healthy bone. OI bone surface shows loosely bound fibers and particles. The appearance of these loosely bound areas demonstrates a weaker interaction between the constituents of OI bone than those in the healthy bone. Some irregular osteons and super big Haversian canal channels are also seen. These big porous structures add to the fragility of OI bone. OI bone also shows abnormal collagen fiber areas hidden in crevices and abnormal depositions of mineral region as a separate cluster. This over mineralized area is filled with globules at the size of several microns. Regular banding pattern is also detected on the normal areas of OI bone specimen. However, the periodicity is measured as about 62.0 nm on average which is smaller than that for the healthy bone (67.0 nm).

Photoacoustic-FTIR (PA-FTIR) spectra of OI and healthy human cortical bone

Fig. 2 shows PA-FTIR spectra for healthy and OI samples from the transverse and longitudinal sections in the energy range of 4000–400 cm^{-1}. The spectra are normalized with the O–H peak (3322 cm^{-1}).

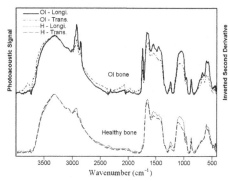

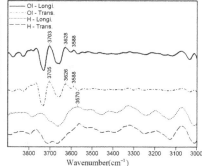

Figure 2. Linear PA-FTIR spectra of human OI cortical bone and healthy cortical bone (longitudinal and transverse sections, respectively) in the 4000-400 cm^{-1} region, at mirror velocity: 0.158 cm/s.

Figure 3. Inverted second-derivative curves in the energy range of 3000-3900 cm^{-1}.

As seen in Fig. 2, significant differences between healthy and OI bone specimen can be observed. Firstly, C-H stretching bands around 2854cm^{-1} and 2926cm^{-1} of OI bone exhibit much higher and sharper peaks than those of healthy bone. These bands are attributed to organic components such as lipids, proteins, carbohydrates and nucleic acids. Similarly, C=O stretching band from lipids, cholesterol esters, triglycerides at 1747cm^{-1} of OI bone also has much higher intensity than that of healthy bone, and the band position remains the same. In addition, a new prominent broad band appears at around 640-710 cm^{-1} in OI bone spectra, which is assigned to C-S stretching vibration[12]. In some cases of type I OI, glycine is replaced by small amino acids, such as alanine, serine and cysteine near the amino terminal ends of the triple-helical domains of either COL1A1 or COL1A2[13]. The appearance of this new band implies that the collagen molecule of this OI tibia may contain more cysteine which possesses C-S bond. The above three differences indicate that OI bone contains more non-collagenous organic components than healthy bone. Additionally, the structure of collagen molecules may possess more cysteine replacements for glycine.

Further, the second-derivative for the 3000-4000 cm^{-1} region is shown in Fig.3. The bands at 3073 cm^{-1} and 3628 cm^{-1} are assigned to asymmetric stretching mode (v3) and symmetric stretching mode (v1) for water associated with HAP. These bands appear as new bands as compared to the healthy bone spectra, indicating that water interacts more closely with HAP in OI bone specimens. The band at 3588 cm^{-1} is assigned to the stretching vibration of the structural hydroxyl group from hydroxyapatite $Ca_{10}(PO4)_6(OH)_2$; however, it is shifted to the higher energy side as compared to that band at 3570 cm^{-1} in the healthy bone spectra, implying that the molecular structure of OI hydroxyapatite is different from the healthy. The bands around 2100 cm^{-1}, OH stretching vibrations from P–OH of OI bone, are more prominent in OI than the healthy bone. The hydroxyapatite present in human bone is calcium (Ca)-deficient and contains specific lattice substitutions such as labile, stable CO_3^{2-} and HPO_4^{2-} species, and ion vacancies in the apatitic crystals as well[14]. Since P-OH originates from HPO_4^{2-}, the stronger appearance of OH stretching vibrations from P-OH of OI bone indicates that OI bone contains more HPO_4^{2-} than the healthy bone.

In addition, as seen from Fig. 2, amide III (C-N stretch, N-H bend, C-C stretch) band of the healthy bone is apparently different from OI bone: a small band at 1275cm^{-1} appears in OI spectra; the peak of amide III in OI spectra is 1241 cm^{-1}, 5 cm^{-1} upshifting from that in the healthy spectra; a band at around 1200 cm^{-1} in the healthy spectra almost disappears in OI spectra. The band of amide III reflects both composition and secondary structure information. Therefore, it is too complex to analyze here. But still, it reflects a significant alteration of organic component of OI bone from the healthy bone.

On comparing the longitudinal section and the transverse section of OI bone, several trends are observed as also seen in the healthy bone: (1) In the longitudinal section, the C-H stretching vibrations at around 2855cm^{-1} and 2926cm^{-1}, and C=O at around 1747cm^{-1} of the longitudinal section are more intense than those of the transverse section; (2) The longitudinal section is also more stoichiometric as analyzed from the curve fitting results in the energy region of 1180-927cm^{-1}. This orientational stoichiometry results from the interaction between the mineral surface and different parts of collagen molecules as discussed in the previous study[6]; (3) The O-H stretching bands from P-OH at around 2100 cm^{-1} from the transverse section is stronger than that of the longitudinal section, indicating that the transverse section contains more HPO_4^{2-} than the longitudinal section. Different from the healthy bone, amide I band (protein C=O stretch, at around 1652cm^{-1}) profiles of the two sections of OI bone are different, exhibiting orientation differences.

In situ FE-SEM nanoindentation

In situ FE-SEM nanoindentation has been performed on OI and the healthy bone samples in the two directions (Fig. 1). The resulting elastic moduli from the transverse sections are shown in Table 1. As seen from Table 1, OI bone has the same trend as the healthy bone: the interstitial lamellae have higher median elastic modulus and hardness than osteonal lamellae. The reason is that the interstitial lamellae are more mature than the osteonal lamellae. The differences between OI and the healthy bone samples are: (1) Median elastic moduli of OI bone are greater than those of healthy bone; (2) Elastic moduli of OI bone have larger range. These differences may be due to the fact that OI bone has higher mineral/collagen ratio[2] and the mineral has abnormal deposition area. Table 2 shows maximum, median, and minimum values of the elastic moduli of OI and the healthy bone samples in the longitudinal section. As seen from the table, the variation of elastic moduli at the lower load is greater than that with higher load in both bone sample types. This phenomenon has been observed previously[15]. Lower load corresponds to shallow indents which reflects more individual constituent properties rather than bulk properties with deep indent. The differences between OI and the healthy bone samples are: (1) The larger variation implies a more heterogeneous feature of OI bone; (2) Median elastic moduli of OI bone are more dependent on the indentation site.

Table 1. Elastic moduli of bone samples in the transverse section

Elastic modulus(GPa)	Healthy osteonal lamellae	Healthy interstitial lamellae	OI osteonal lamellae	OI interstitial lamellae
Min	6.78	15.47	6.51	1.13
Median	18.35	27.92	25.18	30.77
Max	54.28	72.38	138.99	123.51

Table 2. Elastic moduli of bone samples in the longitudinal section

Elastic modulus	Load	Min (GPa)	Median (GPa)	Max (GPa)
	15μN	10.13	32.87	60.90
Healthy bone	50 μN	8.00	17.57	40.20
	100 μN	8.12	10.86	18.82
	15μN	2.96	15.86	172.17
OI bone	50 μN	2.57	15.87	34.04
	100 μN	11.55	27.92	69.72

CONCLUSIONS

In the present study, FE-SEM, PA-FTIR, in situ FE-SEM nanoindentation are utilized to characterize OI (suspected type I) and healthy human cortical bone samples. Infrared spectroscopy studies show that abnormal collagen molecules in OI bone not only affect other organic matrix proteins, but also affect the mineral ion environment. As a result, this altered materials system is suspected to further affect bone remodeling process which causes more porous and fibrous features as well as slightly altered banding pattern of OI bone as shown in

SEM images. SEM images of OI bone also shows loosely attached fibrils and condensed mineral areas as signs of weakened intermolecular adhesion. This heterogeneous condition leads to more local variations in OI bone as demonstrated in the nanoindentation results. As compared to healthy cortical bone, OI bone has similar orientational stoichiometry of hydroxyapatite. In OI, in addition to changes to colleen, mineralization is also influenced and thus also the stoichiometry of the mineralized phase.

ACKNOWLEDGMENTS

Instrumentation obtained from National Science Foundation MRI grants is acknowledged for enabling experiments conducted in this work. Authors would like to acknowledge assistance in electron microscopy laboratory from Mr. Scott Payne.

REFERENCES

1.   R. Dalgleish, Nucleic Acids Research **26** (1), 253-255 (1998).
2.   A. Boyde, R. Travers, F. H. Glorieux and S. J. Jones, Calcified Tissue International **64** (3), 185-190 (1999).
3.   P. Fratzl, O. Paris, K. Klaushofer and W. J. Landis, Journal of Clinical Investigation **97** (2), 396-402 (1996).
4.   N. P. Camacho, W. J. Landis and A. L. Boskey, Connective Tissue Research **35** (1-4), 259-265 (1996).
5.   M. Vanleene, A. Porter, P. V. Guillot, A. Boyde, M. Oyen and S. Shefelbine, Bone **50** (6), 1317-1323 (2012).
6.   C. Gu, D. R. Katti and K. S. Katti, Spectrochimica acta Part A **103**, 25-37 (2013).
7.   S. M. Pradhan, K. S. Katti and D. R. Katti, Journal of engineering mechanics (2012).
8.   D. R. Katti, S. M. Pradhan and K. S. Katti, Journal of Biomechanics **43** (9), 1723-1730 (2010).
9.   R. Bhowmik, K. S. Katti and D. R. Katti, Journal of Materials Science **42** (21), 8795-8803 (2007).
10.   S. M. Pradhan, D. R. Katti and K. S. Katti, Journal of nanomechanics and micromechanics **1** (3), 104-110 (2011).
11.   W. Traub, T. Arad, U. Vetter and S. Weiner, Matrix Biology **14** (4), 337-345 (1994).
12.   G. Socrates, *Infrared and Raman Characteristic group frequencies: tables and charts*, 3rd ed. (John Wiley & Sons, Chichester, 2004).
13.   F. S. van Dijk, P. H. Byers, R. Dalgleish, F. Malfait, A. Maugeri, M. Rohrbach, S. Symoens, E. A. Sistermans and G. Pals, European Journal of Human Genetics **20** (1), 11-19 (2012).
14.   D. Farlay, G. Panczer, C. Rey, P. D. Delmas and G. Boivin, Journal of Bone and Mineral Metabolism **28** (4), 433-445 (2010).
15.   K. S. Katti, B. Mohanty and D. R. Katti, Journal of Materials Research **21** (5), 1237-1242 (2006).

Engineering and Application of Bioinspired Structured Materials

Mater. Res. Soc. Symp. Proc. Vol. 1621 © 2014 Materials Research Society
DOI: 10.1557/opl.2014.64

Mechanosensitive Channels Activity in a Droplet Interface Bilayer System

Joseph Najem[1], Myles Dunlap[1], Sergei Sukharev[2], and Donald J. Leo[3]

[1]Biomolecular Materials and Systems Laboratory, Virginia Tech, Blacksburg, VA 24061, U.S.A.
[2]Department of Biology, University of Maryland, College Park, MD 20742, U.S.A.
[3]College of Engineering, University of Georgia, Athens, GA 30609, U.S.A.

ABSTRACT

This paper presents the first attempts to study the large conductance mechano-sensitive channel (MscL) activity in an artificial droplet interface bilayer (DIB) system. A novel and simple technique is developed to characterize the behavior of an artificial lipid bilayer interface containing mechano-sensitive (MS) channels. The experimental setup is assembled on an inverted microscope and consists of two micropipettes filled with PEG-DMA hydrogel and containing Ag/AgCl wires, a cylindrical oil reservoir glued on top of a thin acrylic sheet, and a piezoelectric oscillator actuator. By using this technique, dynamic tension can be applied by oscillating axial motion of one droplet, producing deformation of both droplets and area changes of the DIB interface. The tension in the artificial membrane will cause the MS channels to gate, resulting in an increase in the conductance levels of the membrane. The results show that the MS channels are able to gate under an applied dynamic tension. Moreover, it can be concluded that the response of channel activity to mechanical stimuli is voltage-dependent and highly related to the frequency and amplitude of oscillations.

INTRODUCTION

Biomolecular unit cells can be described as small building blocks whose repetition can form the basis of a novel biomolecular material system[1]. The biomolecular unit cell consists of a lipid bilayer interface formed at the contact of two aqueous droplets encased in lipid monolayers. The droplets are surrounded by a hydrophobic organic solvent (Hexadecane), and are sitting on fixed silver-silver chloride (Ag/AgCl) electrodes[2,3]. Many types of biomolecules, such as ion channels can self-assemble in the lipid bilayer interface. Therefore, the Droplet Interface Bilayer (DIB) has been extensively used to systematically study the activity of various biomolecules including alamethicin[4], bacteriorhodopsin[5], and many more. Other types of biomolecules such as mechanosensitive (MS) channels self-assembled within the DIB should be able to respond to an expansion in the artificial membrane. Therefore, it can be used as a model system to understand how the structure of the protein and its incorporation into the unit cell affects its transduction properties. MS channels residing in the cytoplasmic membrane of Escherichia coli respond to a mechanical tension in the cell membrane[6], and fall under three categories according to their conductance level[7]. MscL, the mechanosensitive channel of large conductance, has been studied both in vivo and in vitro using patch-clamp methods[8].

In this paper, the incorporation and activation of MscL is investigated. These channels, usually found in E-Coli bacteria, respond to a mechanical tension in the cell membrane by either opening or closing. Hence, mechanosensitive channels self-assembled within the lipid bilayer interface should be able to respond to any change in the artificial membrane tension. However, when membrane is under tension, excess lipids existing in the aqueous phase will self-insert in the membrane, leading to tension relaxation. Therefore, the MscL channels incorporated within the lipid bilayer are tested by applying a dynamic oscillation of the membrane rather than static. The application of a dynamic oscillation will result in stretching the lipid bilayer membrane at low frequencies of oscillations, where other vibrational modes will appear at high frequencies[9, 10]. This phenomenon is also shown and proved in this paper through a series of mechanoelectrical response experiment of the artificial lipid bilayer membrane.

Two different types of MS channels are studied in this paper: (1) MscL, a nonselective MS channel with large conductance (3-nS, in 200 mM KCl), and is activated at relatively high tensions (~14 dynes/cm)[6, 11]. In addition, (2) MscL V23T mutant, which is a gain-of-function mutant of MscL[12]. V23T mutant activates at lower tension similar to the MS channel with small conductance (MscS), which is around 1.8 times lower than MscL (~7.8 dynes/cm)[12].

EXPERIMENTAL METHODS AND MATERIALS

The technique described in this report has been developed to characterize the behavior of artificial lipid bilayer interface containing mechanosensitive (MS) channels. Using this technique, dynamic tension can be applied axially to the droplet interface bilayer through squeezing and releasing the droplets. The experimental setup is centered on an inverted microscope (AxioSkop-ZEISS), and consists of two micropipettes filled with PEG-DMA hydrogel, and also contains Ag/AgCl wires, a cylindrical oil reservoir glued on top of a thin acrylic sheet, and finally, a piezoelectric oscillator (Figure 1).

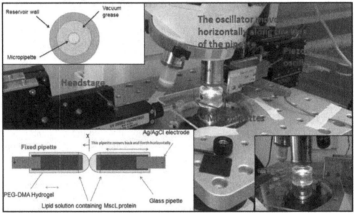

Figure 1: The experimental setup which includes the headstage, the micropipette holder, and the piezoelectric oscillator.

The flat tip micropipettes are made of borosilicate glass with 1 mm and 0.5 mm outer and inner diameters respectively. The micropipettes are initially filled with a UV curable hydrogel, and then a silver/silver-chloride (Ag/AgCl) wire is fed into the micropipette through the hydrogel. The hydrogel is then cured through free-radical photopolymerization upon exposure to UV light for 3 min at 1 W intensity, 365 nm UV source. The hydrogel solution with a concentration of 40 % (w/v) PEG-DMA contains 0.5 % (w/v) Irgacure 2959, and is mixed with a 500 mM KCl and

10 mM MOPS, pH 7 electrolyte solution. A spot UV source (LED-100, Electro-Lie Corp) is used, as well as Poly(ethylene glycol) dimethacrylate polymer (PEG-DMA; MW=1000 g/mol). Irgacure 2959 is used as a photoinitiator in this study and was obtained from Ciba.

The micropipettes are then placed horizontally in opposite directions in a custom-made apparatus. The apparatus is made by gluing an acrylic cylinder on a thin acrylic sheet using epoxy. Two opposing holes (1.1 mm in diameter) are drilled on the wall of the cylinder at 1 cm from the bottom. Two other holes, concentric to the previously drilled holes with a diameter of 4 mm, are also drilled. In order to prevent hexadecane oil from leaking, vacuum grease is deposited around the glass micropipettes. Vacuum grease is used because it is soft; hence no vibration will be transmitted from the oscillator to the oil reservoir and consequently to the micropipette connected to the positive lead.

Materials

The aqueous phases that form the droplets consist of a suspension of phospholipids vesicles and a buffering agent in highly pure deionized water, while the oil phase consists of Hexadecane (99%, Sigma). The lipid vesicle solution is prepared and stored as described in many articles previously published[2, 13]. The lipids solution contains 2mg/ml solution of 1,2-diphytanoyl-sn-glycero-3-phosphocholine (DPHPC, Avanti Polar Lipids, Inc.) vesicles in 500mM potassium chloride (KCl, Sigma), 10mM 3-(N-morpholino)propanesulfonic acid (MOPS, Sigma), pH7. MscL and V23T mutant are received form Dr. Sergei Sukharev's lab. The received material is initially reconstituted in DPhPC liposomes and then diluted in the DPhPC-lipid vesicle solution to yield a final concentration around 0.2μg/ml.

Bilayer formation and Electrical recordings

Lipid bilayer interface formed within the biomolecular unit cell is characterized through two types of electrical measurements. Electrically, the lipid bilayer interface is modeled as a capacitor and a resistor in parallel. Therefore, capacitance measurements are carried out in order to verify the increase in capacitance resulting from the bilayer formation. Axopatch 200B and Digidata 1440A (Molecular Devices) are used to measure the resulting square-wave current produced by an external, 10 mV triangular voltage waveform at 10 Hz (Hewlett Packard 3314A function generator). The second type of electrical recording is a current measurement of the bilayer interface, which is held under voltage-clamp while mechanically oscillating the bilayer containing the MS channels. All electrical recordings are carried out under a lab-made Faraday cage that serves as an electrical shield.

RESULTS AND DISCUSSION

Droplet interface bilayer mechanoelectrical response

The DIB mechanoelectrical response was observed at a variety of frequencies by using a linear sinusoidal-sweep (chirp signal) from the piezoelectric actuator, as well as single frequency sinusoidal oscillations. Figure 2(a) shows the response of the bilayer before and after the application of the mechanical oscillations while an external 10 Hz triangular voltage signal is applied. When applying the mechanical excitation, the current response exhibits a change in the amplitude as a result of the change in capacitance, which reflects the zipping and unzipping of the bilayer. This observation is clearly portrayed in Figure 2(b), where no external voltage signal is applied and the current variation is only related to the change in capacitance (dC/dt). This experiment served not only to understand the mechanoelectrical response of the DIB, but also as a baseline for the mechanosensitive channels' experiments. A DC potential with amplitudes ranging from -150 mV to 150 mV is applied, resulting a highlight two major observations: (1) the magnitude of the current does increase proportionally with the applied voltage, and (2) no traces or any gating-like event happened as shown in Figure 2(b).

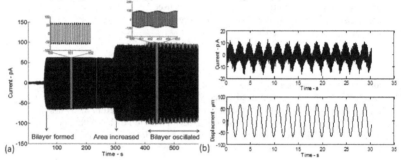

Figure 2: (a) Bilayer formation and current amplitude change due to membrane oscillation (No protein), (b) Current response of the bilayer (No protein) due to a change in bilayer are resulting from membrane oscillation.

Effect of protein concentration on the bilayer formation

When reconstituting MscL and V23T mutant in the liposome, it is noticed that the bilayer formation is inhibited as shown in Figure 3(a). Specifically, the bilayer containing MscL when "formed" did not exhibit a significant increase in capacitive current (~15 pA peak-to-peak). The current response displays a triangular shape waveform which reflects a leaky bilayer (i.e. low resistance bilayer). In the course of trying to understand what is leading to such a behavior, it is found that the protein to lipid ratio in our solutions is at least three orders of magnitudes higher than the values reported in the literature[14]. Therefore, the protein to lipid ratio is reduced by diluting the solutions with DPhPC lipid solution. The formed bilayer seemed more consistent with previous bilayer formation results when no MscL protein existed in the bilayer. The resulting capacitive current (~ 200 Pa) significantly increased upon thinning of the bilayer, and exhibits a square-shape like waveform, which reflects a high resistance bilayer (Figure 3(b)). Note that in this paper the optimum protein concentration is not explored. The concentration is reduced in a manner just to be able to form a functional bilayer to perform the experiments.

Mechanosensitive channels activation

The diluted solutions containing MscL V23T mutant are used to form artificial lipid bilayer membranes. The membranes are then mechanically oscillated at different frequencies ranging from 0.5 Hz to 75 Hz using sinusoidal and step waveforms. Simultaneously, a DC potential

174

ranging from 0 to 100 mV is applied. The results show that sub-conductance gating events are taking place at the peak of the mechanical oscillation, which corresponds to the larger bilayer formed. These events occurred when a DC potential higher than 80 mV is applied. Three important observations are made from Figures 4 (a) and (b): (1) The gating events happened at the maximum compression of the bilayer. This may be an indication that the membrane is mechanically stretched, leading to channel gating. (2) It is shown that the gating frequency and conductance levels of the V23T mutant are related to the applied DC potential across the membrane. Figure 4 (b) shows what is considered full gating (around second 75) when a potential of 100 mV, which is higher than the one applied in Figure 4(a). Moreover, some sub-conductance gating event were seen even when the mechanical oscillations are turned off and a DC potential is applied, which emphasizes the role of applied potential on the gating of MS channels. (3) While running the experiments, the importance of the frequency of oscillations was obvious. Mainly, the gating events happened at lower oscillations frequencies and disappeared while the frequency is increased. As a result, it is believed that most of the events seen are sub-conductance states of the V23T mutant.

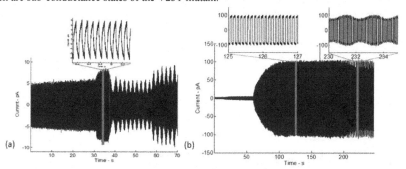

Figure 3: (a) Bilayer formation inhibited due to high protein concentration (MscL V23T mutant), (b) Bilayer formed normally after reducing the protein (MscL V23T mutant) concentration.

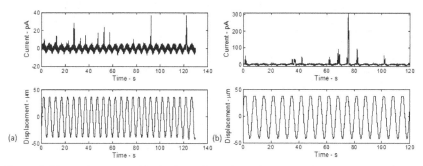

Figure 4: The results presented in the these plots refer to the MscL V23T mutant protein (a) Bilayer current response when a 0.2 Hz sinusoidal waveform in applied along with a 90 mV DC

175

potential, (b) Bilayer current response when a 0.2 Hz sinusoidal waveform in applied along with a 100 mV DC potential.

CONCLUSIONS

The results presented in this paper highlight the first gating events of MS channels in the DIB system. The MS channels incorporated in the lipid bilayer responded to an applied mechnical oscillation by opening and increasing the conductance level of the membrane. All experiments are conducted using a novel experimental setup that features two aqueous droplets placed at the tip of two horizontally opposing hydrogel filled micropippettes. As expected, it was revealed that the mechanical tension has a major contribution to the gating of the MS channels. In the end, it can be concluded that the response of channel activity to mechanical stimuli is voltage-dependent and is highly related to the frequency and amplitude of oscillations.

ACKNOWLEDGMENTS

The authors gratefully acknowledge financial support through the Air Force Office of Scientific Research Basic Research Initiative grant number 11157642.

REFERENCES

1. S. A. Sarles, L. J. Stiltner, C. B. Williams and D. J. Leo, ACS applied materials & interfaces 2 (12), 3654-3663 (2010).
2. S. A. Sarles and D. J. Leo, Analytical chemistry 82 (3), 959-966 (2010).
3. S. A. Sarles, Virginia Polytechnic Institute and State University, 2010.
4. S. A. Sarles and D. J. Leo, Journal of intelligent material systems and structures 20 (10), 1233-1247 (2009).
5. W. L. Hwang, M. Chen, B. Cronin, M. A. Holden and H. Bayley, Journal of the American Chemical Society 130 (18), 5878-5879 (2008).
6. S. I. Sukharev, B. Martinac, V. Y. Arshavsky and C. Kung, Biophysical Journal 65 (1), 177-183 (1993).
7. P. Blount, S. I. Sukharev, P. C. Moe, B. Martinac and C. Kung, in *Methods in Enzymology*, edited by P. M. Conn (Academic Press, 1999), Vol. Volume 294, pp. 458-482.
8. S. I. Sukharev, B. Martinac, P. Blount and C. Kung, Methods 6 (1), 51-59 (1994).
9. A. G. Petrov, Analytica Chimica Acta 568 (1–2), 70-83 (2006).
10. S. A. Sarles, J. D. Madden and D. J. Leo, Soft Matter 7 (10), 4644-4653 (2011).
11. S. I. Sukharev, W. J. Sigurdson, C. Kung and F. Sachs, The Journal of General Physiology 113 (4), 525-540 (1999).
12. B. Akitake, A. Anishkin and S. Sukharev, The Journal of General Physiology 125 (2), 143-154 (2005).
13. G. Szabo, M. C. Gray and E. L. Hewlett, Journal of Biological Chemistry 269 (36), 22496-22499 (1994).
14. N. Saint, J.-J. Lacapère, L.-Q. Gu, A. Ghazi, B. Martinac and J.-L. Rigaud, Journal of Biological Chemistry 273 (24), 14667-14670 (1998).

Mater. Res. Soc. Symp. Proc. Vol. 1621 © 2014 Materials Research Society
DOI: 10.1557/opl.2014.1

Cellulose Nanofibril (CNF) Reinforced Starch Insulating Foams

N. Yildirim[1,2], S.M. Shaler[1,2], D.J. Gardner[1,2], R. Rice[2] , D.W. Bousfield[3]

1. The Advanced Structures and Composites Center, University of Maine, 35 Flagstaff Road, Orono, ME, 04469-5793 T: (207) 581-2123, F: (207) 581-2074

2. School of Forest Resources, University of Maine, 5755 Nutting Hall, Orono, ME 04469-5755 T: (207) 581-2841, F: (207) 581-2875

3. Department of Chemical and Biological Engineering, University of Maine, 5737 Jenness Hall, Orono, ME 04469-5737 T: (207) 581-2277, F: (207) 581-2323

ABSTRACT

In this study, biodegradable foams were produced using cellulose nanofibrils (CNFs) and starch (S). The availability of high volumes of CNFs at lower costs is rapidly progressing with advances in pilot-scale and commercial facilities. The foams were produced using a freeze-drying process with CNF/S water suspensions ranging from 1 to 7.5 wt. % solids content. Microscopic evaluation showed that the foams have a microcellular structure and that the foam walls are covered with CNF`s. The CNF's had diameters ranging from 30 nm to 100 nm. Pore sizes within the foam walls ranged from 20 nm to 100 nm. The materials` densities ranging from 0.012 to 0.082 g/cm^3 with corresponding porosities between 93.46% and 99.10%. Thermal conductivity ranged from 0.041 to 0.054 W/m-K. The mechanical performance of the foams produced from the starch control was extremely low and the material was very friable. The addition of CNF's to starch was required to produce foams, which exhibited structural integrity. The mechanical properties of materials were positively correlated with solids content and CNF/S ratios. The mechanical and thermal properties for the foams produced in this study appear promising for applications such as insulation and packaging.

INTRODUCTION

In this study cellulose nanofibril and starch foams were produced and characterized. The main reason for using cellulose is that it is an abundant material, which can be obtained from renewable sources including a broad range of plants and sea animals (tunicates) [1]. Starch is another abundant natural polymer, which is a promising raw material for the development of novel materials [2]. It is a widely available biopolymer with a price half that of polyethylene and polystyrene. The mechanical properties of starch are highly correlated with density and amylose content. Increasing the solids content of starch suspensions increases the starch paste viscosity, which decreases the rate of steam bubble expansion, resulting in higher densities and mechanical properties. Normal cornstarch has higher elastic modulus (E=220 MPa) than wheat, potato and tapioca starches [3]. Annually, millions of metric tons of starch are used as non-food

177

products in the paper and textile industries [4]. Starch, which has two major components, amylose consisting of α-(1-4)-linked D-glucose and amylopectin with a myriad α-(1-6)-linked branch point, is not a good choice as a replacement for any plastic because it is mostly water soluble, difficult to process and has low mechanical properties. It was found that reinforcing starch with cellulose microfibrils increases the mechanical properties significantly [5]. Glenn and Irving produced microcellular starch foams with different drying techniques and investigated the mechanical and thermal properties. They showed that mechanical properties of microcellular foams are positively corrected with density. They found that corn starch foams exhibited greater compressive strength (0.19 -1.14 MPa) and density (0.12-0.31 g/cm^3) than the wheat starch and high amylose cornstarch foams. Corn starch foams have been made with thermal conductivity values ranging from 0.037-0.040 W/m-K [6]. Tatarka and Cunningham compared eight commercial starches and expanded polystyrene (EPS)-based loose-fill foam products. They showed that starch based foams have two or three times higher density (0.0216 g/cc) than the EPS foams. Most starch-based foams have similar compressive strength (0.0927 MPa) with EPS foams [7]. Chen et al., studied starch graft poly (methyl acrylate) loose-fill foams and their properties [8]. The starch-based foams were prepared using graft polymerization. They found that starch graft poly (methyl acrylate) foams (S-g-PMA foams) have 0.07±0.01 MPa compressive strength with 0.0086±0.00021 g/cm^3 density [8]. Nabar et al., showed that starch based foams have compressive strength between 12.5-13.1 Pa with the densities changes from 0.003 to 0.0035 (g/cm^3) [9] . Bhatnagar and Hanna [10] investigated the plastic foams produced from various starch sources using extrusion. The corn, tapioca, wheat, rice and potato starches were extruded with 30% polystyrene and 5 % magnesium silicate or 1% polycarbonate or 0.5% azodicarbonamide in a single screw laboratory extruder. They showed that starch based foams can be prepared from different starch sources by replacing 70% polystyrene with starch. The properties of the end product depend on the additives and source of starch. The polycarbonate and magnesium silicate gave the best result for all the starches [10]. Svagan et. al. investigated the mechanical properties of amylopectin-based foams with varying microfibrillated cellulose (MFC) contents and they showed increasing the MFC content produces higher mechanical properties [11]. Starch needs to be modified due to its low mechanical properties [12]. Adding, PLA (poly lactic-acid) and PHEE (poly hydroxyester ether) decreases starch foam density by increasing the radial expansion, the ratios of the cross-sectional area of the foam to the die cross-section. Friability is the fragmentation of foam during handling. Adding resin to starch foams reduces the friability. Denser foams typically exhibit thicker cell walls and hence produce denser foams, which have better resistance to deformation than lower density foams with thinner cell walls [13]. Thermoplastic starch is starch, which has been modified using plasticizers (water, glycerin, sorbitol etc.) at high temperatures (90-180 °C) under shear. This type of starch has two disadvantages; water solubility and low mechanical properties. Mixing natural fibers with polysaccharides like thermoplastic starch significantly improves the mechanical properties. The addition of 5-10% aspen wood fiber has been shown to significantly increase the mechanical properties [12]. Adding monostearyl citrate (MSC) gives significant improvement in water resistance [12]. Adding 16% of fiber by weight to the matrix was shown to increase the tensile strength 156 %. Adding fiber decreased the moisture sorption. In addition, it was determined that starch is more hydrophilic than cellulose [14]. Reinforcing the starch with α-cellulose, softwood fiber, hardwood fiber or municipal solid waste fiber increases the tensile strength. This exploratory study aimed to determine the impact of solids content and CNF/S ratio on the morphology, physical, and mechanical properties of foams with the intent to evaluate their suitability for application as structural insulation foam or other market opportunities.

EXPERIMENTAL DETAILS

In this study, thermal insulation foams were produced by using biodegradable polymers that have low thermal conductivities with satisfactory mechanical properties. Five different thermal insulation foams were prepared from aqueous suspensions (tap water+material) with the following solid contents; 1% CNF, .5% CNF+.5% starch, 1.5% CNF+3% starch, 1.5% CNF+6% starch and 7.5% starch. The suspension was placed in a 20 L capacity container. A high shear mixer was used to disperse the CNF in suspension (1700 RPM for 20 minutes). For creating the starch foams, industrial corn starch (Tate&Lyle) was used and cooked at 87.8 °C (190 °F) and mixed at 500 RPM for one hour. Starch solutions were cooled down to the room temperature (23±2°C). The final solids content was determined by oven drying 30-35 gram suspension samples. The suspensions had a high consistency and were gel-like in appearance. Starch solution and CNF suspension were put into high shear mixer and dispersed (1700 RPM for 20 minutes). The dispersed suspensions were poured into trays (30.6 cm x 61 cm) to a depth of 3.5 cm and placed in a freeze dryer. Suspensions were freeze-dried using a Millrock Technology Max53 freeze dryer utilizing the Opti-Dry 2009 control system. T-type thermocouples were placed in the material to monitor temperature during the freeze-drying process. A partial vacuum was then pulled to exclude ambient moisture from entering the freeze drying chamber. The chamber temperature was lowered from 20 °C to -45°C in 1 hour and maintained at that temperature for 250 minutes. The chamber was then evacuated to a pressure of 100 mTorr. The chamber temperature was maintained at -45°C for 30 minutes, ramped to 0°C over 2 hours, ramped to 20°C in 4 hours and then maintained until average thermocouple reading in the materials was 20°C for 4 hours.

Morphological, physical, mechanical and thermal properties of insulation foams were investigated. Representative CNF/starch foam samples were imaged using scanning electron microscopy (SEM) and atomic force microscopy (AFM). Density measurements of the foams were performed according to ASTM C303-10 by measuring six 150 mm x 150 mm x 25.4 mm (6 in x 6 in x 1 in) specimens from each group. The void fraction, which is called porosity, the ratio of pore volume to its total volume of foams, was calculated using liquid porosimetry method [15]. Six (6) samples with 300 mm x 100 mm x 25.4 mm (12 in x 4 in x 1 in) dimensions from each sample group were tested by using a three-point bending test method according to ASTM C203-12. The crosshead displacement rate was 6 mm per minute. Specimen displacement was obtained from the crosshead displacement (Instron 5966, with 100KN maximum load). Flexural tests were applied under laboratory conditions (25±2 °C and 50% relative humidity). The flexural modulus of the foams was obtained from the linear initial part of the force-deflection curves (Figure 1).

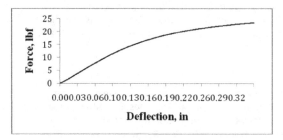

Figure 1. Typical force-deflection curve for flexural tests of foams.

Six (6) samples with 150 mm x 150 mm x 25.4 mm (6 in x 6 in x 1 in) dimensions from each sample group were tested according to ASTM C165-07. Each specimen was compressed at a rate of 6 mm per minute and the specimen displacement was obtained from the crosshead displacement (Instron 5966, with 100KN maximum load). The compression modulus was obtained from the linear initial part of the force-deflection curves (Figure 2).

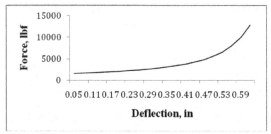

Figure 2. Typical force-deflection curve for compression tests of foams.

Six (6) specimens (300 mm x 300 mm x 25.4 mm) were prepared from each group, two specimens from each tray. The steady-state thermal transmission was measured according to ASTM C518-10 using a heat flow meter (NETZSCH Lambda 2000 heat flow meter). Six (6) specimens from each group, which weighed from 6 g to 10 g, were prepared as a powder and placed into crucibles according to ASTM E113-08. The thermogravimetric analysis was conducted from 25 °C to 800 °C with 10 °C increase per minute using a Mettler Toledo Thermogravimetric Analyzer (TGA/SDTA851[e]).

DISCUSSION

The large range in sample density (0.013 to 0.098 g/cm3) resulted in all statistical analysis having significant overall differences. There was a significant effect of solids content on the porosity and density (Table 1). As expected, including less solids content in the suspension produced a more porous structure. The reason for the inverse proportion between solid content and porosity can be explained by the increased bonds with increase solid content between starch and CNF in the same volume. This indicates that the production process (suspension-dispersion) and foam preparation method (freeze-drying) can be manipulated to produce foams of varying density and porosity.

Table 1. Porosity properties of foams

Sample	Solid-Water Content	Density (g/cm^3)	Porosity (%)
0.5% CNF+ 0.5%Starch	1%-99%	0.013 (0.0010) D	99.10 (0.081) A
1% CNF	1%-99%	0.014 (0.0016) D	99.10 (0.118) A
1.5% CNF+ 3%Starch	4.5%-95.5%	0.053 (0.0011) C	96.50 (0.877) B
1.5% CNF+ 6%Starch	7.5%-92.5%	0.076 (0.0020) B	94.95 (0.131) C
7.5% Starch	7.5 %-92.5%	0.098 (0.0036) A	93.46 (0.231) D

Parentheses indicate the standard deviation. A,B,C and D letters indicates the significant differences between the treatments.

Fiber size measurements were obtained from SEM images. A representative set of SEM images and fiber size measurements of the foams are given in Figure 3. Some damage (cracking) was evident because of sample preparation process (razor blade cutting). The structure of the foam wall material was evaluated (Figure 3b). The foam wall structure is a plate of CNF material embedded within a starch matrix. Further investigation of the foam cell wall material (Figure 3c) illustrates the nanoscale fibril structure of the CNF with diameters ranging from 30 to 100 nanometers.

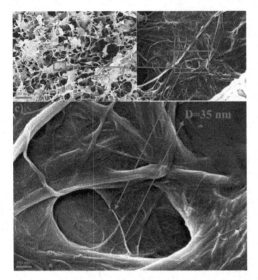

Figure 3. Representative SEM images of foams (1.5%CNF+6%S), (a) cellular structure, (b) distribution of nanofibrils in 2 micron scale bar and (c) fibril diameter measurements.

Atomic force microscopy (AFM) was used to measure the diameter of nano pores in the cellular wall material. It was determined that the diameter of the pores (Figure 4) ranged from 20 nm to 100 nm. The difference between the SEM and AFM images can be explained by the difference in imaging principles and different regions and field of view of the images.

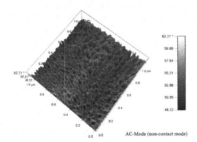

Figure 4. Representative AFM images of foams (1.5%CNF+6%S)

The density, flexural modulus and flexural strength (MOR) of the foams are summarized in Table 2. Density of the foams ranged between 0.013-0.082 g/cm^3. The increase in the solids content increased the total mass of the foam structure, which increased the density. The elastic modulus (MOE) and modulus of rupture (MOR) plotted against density are shown in Figure 5.

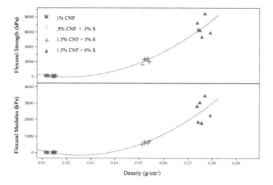

Figure 5. The flexural modulus and flexural strength of foams plotted against density

Increasing the density was correlated with increases in modulus of elasticity (MOE) and modulus of rupture (MOR) and a curvilinear relationship was indicated between density and flexural properties. However, pure starch foams with even higher solid content (7.5%) and density exhibited very low mechanical properties. Reinforcement of the starch foams using CNF produced significant increases in flexural properties for all groups. However, when the 1% CNF (star sign in Figure 5) and .5% CNF + .5% S (rectangular sign in Figure 5) combinations were compared, it seems there is no effect of adding CNF because of the given scale in "y" direction. The CNF addition produced an 84% increase in MOE and 71% increase in MOR (Table 2). In addition to this, when 3% starch or 6% starch was reinforced with 1.5% cellulose nanofibrils, flexural properties increased significantly.

Table 2. Flexural properties of foams

Sample	Density (g/cm^3)	Elastic Modulus (kPa)	Modulus of Rupture (kPa)
.5% CNF+ .5%Starch	0.013 (0.0007) D	2.30 (1.10) C	40.00 (10.0) C
1% CNF	0.014 (0.0012) D	14.0 (8.00) C	140.0 (40.0) C
1.5% CNF+ 3%Starch	0.053 (0.0011) C	610 (80.0) B	2140 (300) B
1.5% CNF+ 6%Starch	0.082 (0.0049) B	2530 (670) A	6590 (1110) A
7.5% Starch	0.098 (0.0036) A	N/A	N/A

Parentheses indicate the standard deviation. A, B, C and D letters indicates the significant differences between the treatments. The extreme friability of the 7.5% starch specimens did not allow for determination of the mechanical performance.

Shey et al. indicated that the commercial polystyrene structural foam (Styrofoam™) has a 1610 kPa stress at yield [16]. Gypsum board has 4600 kPa modulus of rupture in the machine direction and 1500 kPa modulus of rupture in cross direction [17]. The performance of the 1.5% CNF/6% starch formulation exceeded both levels of performance.

The compression performance of the foams (Table 3) showed a trend consistent with that found for flexural behaviour. Density was positively correlated with compression modulus, compressive resistance and the maximum force when the specimens were compressed to 10% of specimen thickness (Figure 6).

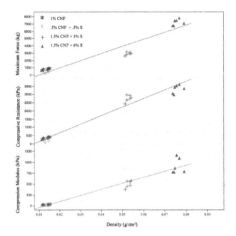

Figure 6. Maximum force applied when the foams were compressed up to 10% of original thickness, compression modulus and compression resistance plotted against density

The role of starch and CNF was directly evaluated at 1% solids content (1% CNF vs. 0.5% CNF + 0.5% Starch). The foam made with half starch resulted in significant decreases in compression modulus (50%), compressive resistance (60%) and maximum force (59%). For a constant solids content (1.5%) of CNF, increasing the starch content in the structure from 3% to 6% increase the compression modulus (83%), compressive resistance (21%), and maximum force (148%).

Table 3. Compression properties of foams and comparisons with other studies

Sample	Density	Compression	Compressive	Fmax at 10% (kg)
.5% CNF+ .5%Starch	0.014 (0.0010) C	15.0 (2.0) C	144 (44.0) C	326.8 (98.80) D
1% CNF	0.014 (0.0016) C	30.0 (7.0) C	359 (63.0) C	792.8 (108.90) C
1.5% CNF+ 3%Starch	0.053 (0.0011) B	496 (82) B	2753 (203) B	2897.3 (213.6) B
1.5% CNF+ 6%Starch	0.076 (0.0020) A	907 (171) A	3330 (249) A	7195.7 (429.0) A
7.5% Starch	N/A	N/A	N/A	N/A
Carbon foam[18]	0.73	-	6100	-
Gypsum board[17]	-	-	2750	-
CNF foam[19]	0.016	62	-	-
CNF foam[19]	0.027	249	-	-
CNF foam[19]	0.063	1760	-	-
CNF aerogel[20]	0.014	34.9	3.20	-
CNF aerogel[20]	0.029	199	24.4	-
CNF aerogel[20]	0.050	1030	69	-
CNF aerogel[20]	0.105	2800	238	-

Parentheses indicate the standard deviation. A, B, C and D letters indicates the significant differences between the treatments.

The comparison of foam compression properties (Table 3) showed that the compression results for the foams produced in this study appear promising for applications such as insulation and packaging. Ali and Gibson [19] found that, the 2 wt. % CNF foams have 1760 kPa compression modulus with 0.063 g/cm^3 density, whereas foams produced in this study (1.5% CNF+6% S) have 907 kPa compression modulus with 0.076 g/cm^3 density. Including 6% starch, and less CNF content in the foam structure produced lower mechanical properties. H. Sehaqui et al. [20] reported that CNF aerogel were positively correlated with density. The reinforced starch foams, which were produced in this study, have similar, comparable compression properties with pure CNF foams, aerogels and some other insulation foams.Thermal conductivity and resistivity results (Table 4) indicated no statistically significant differences among the foam combinations, which have 4.5% or higher solid content in its structure. The pure CNF with 1% solid content showed higher thermal conductivity when compared the other combinations where the lower thermal conductivity means better insulation.

Table 4. Thermal conductivity and thermal resistivity properties of foams

Sample	Density (g/cm³)	Thermal Conductivity (W/m-	R value
.5% CNF+ .5%Starch	0.013 (0.0007) D	0.048 (0.003) B	3.03 (0.34) B
1% CNF	0.014 (0.0012) D	0.054 (0.004) C	2.87 (0.37) B
1.5% CNF+ 3%Starch	0.053 (0.0011) C	0.047 (0.002) AB	3.60 (0.16) A
1.5% CNF+ 6%Starch	0.082 (0.0049) B	0.042 (0.001) A	4.02 (0.06) A
7.5% Starch	0.098 (0.0036) A	0.041 (0.002) A	4.14 (0.18) A
Gypsum Board[19]	-	-	0.83
Silica Aerogels[21]	-	0.041-0.098	-
Fiberglass-	-	0.021	-
Fiberglass-rigid[22]	-	0.330	-
Urethane-rigid[22]	-	0.024	-
Perlite[22]	-	0.054	-
Extruded	-	0.029	-
Urethane (roof-	-	0.024	-
Granular Starch[23]	-	0.490	-
Gelayinized Starch[23]	-	0.470	-
Freeze-dried corn starch[24]	-	0.040	-

Parentheses indicate the standard deviation. A, B and C letters indicates the significant differences between the treatments.

The comparison of thermal insulation properties (Table 4) showed that the thermal conductivity and thermal resistivity results for the foams produced in this study appear promising like the compression properties. Foams produced in this study have 3 to 4 times better thermal resistivity properties when compared to gypsum board that has 0.83 °F.h.ft²/BTU R-value (thermal resistivity) for 25.4 mm thickness [17] and the foams have similar thermal conductivity with nanoporous silica aerogel impregnated highly porous zirconia ceramics have thermal conductivity from 0.041 W/m-K to 0.098 W/m-K [23]. Mahlia et al. showed that, thermal conductivities of some insulation materials as follows: fiberglass-urethane 0.021 W/m-K, fiberglass-rigid 0.33 W/m-K, urethane-rigid 0.024 W/m-K, perlite 0.054 W/m-K, extruded polystyrene 0.029 W/m-K and the urethane (roof deck) 0.021 W/m-K [24]. Hsu and Heldman showed that, granular starch has 0.49 W/m-K thermal conductivity and gelatinized starch have 0.47 W/m-K thermal conductivity [23], which are so conductive when compared the CNF reinforced starch foams produced in this study. Glenn and Irving showed that, corn freeze-dried starch has a 0.040 W/m-K thermal conductivity [24], which is almost same when compared to our study due to similar production method and cellular structure of materials. As a result of thermal analyses (Table 5), it was found 1% CNF foams have a higher thermal degradation point of (onset temperature) 277 °C when compared to the other combinations tested. Petersson et al., indicated that the onset temperature of cellulose nanoparticles is between 200-300 (°C) [23] . The addition of starch consistently decreased onset temperature with reduction from 277 to 255 °C as initial starch concentration increased to 6%. Switching the 0.5% CNF to 0.5% starch or adding more starch to the foam, decreased the onset temperature.

Table 5. TGA and DTGA results for foams

Sample	T, Weight loss 10%, °C	T, Weight loss 50%, °C	DTGA temp. °C	Mass loss (%)	Residue (%)
.5% CNF+ .5%Starch	276 (1.08) AB	320 (2.35) B	303 D	33.2 (1.74) D	14.4 (2.73) A
1% CNF	277 (6.16) A	335 (0.41) A	339 A	55.3 (0.69) A	15.3 (1.32) A
1.5% CNF+ 3%Starch	260 (3.68) ABC	308 (1.17) C	304 C	45.5 (0.71) B	8.70 (1.56) C
1.5% CNF+ 6%Starch	255 (12.0) C	318 (1.09) B	310 B	40.9 (1.40) C	11.6 (1.53) B
7.5% Starch	259 (1.91) BC	300 (0.35) D	295 E	43.3 (0.58) BC	14.9 (0.49) A

Parentheses indicate the standard deviation. A, B and C letters indicates the significant differences between the treatments.

The DTGA (derivative TGA) temperature (decomposition temperature) was determined to be 338.6 °C for 1% CNF. When the starch was added to the structure, two peaks were evident in the DTGA curves, one for starch and one for CNF (Figure 7b). DTGA temperatures were determined statistically different for CNF + starch combinations, whereas they showed similar results that changes between 303 °C to 310 °C (Table 5).

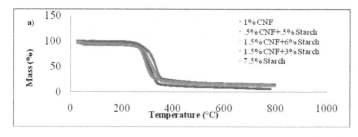

187

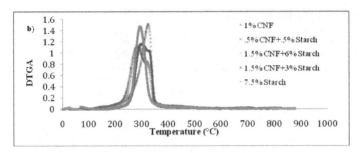

Figure 7. TGA and DTGA curves of foams, (a) TGA curves, (b) DTGA curves

CONCLUSIONS

Biodegradable and renewable foams produced in this study have highly porous microcellular structure includes foam walls of CNF material embedded within a starch matrix. Due to starch`s low mechanical properties cellulose nanofibrils were used as a reinforce material which provided superior performance to starch foams. Increasing the CNF amount and increasing the density produced higher mechanical properties. The optimum properties were obtained from 6% Starch + 1.5% CNF combinations, which has promising mechanical and thermal properties when compared with previous studies. This study was an exploratory and that it has found promising results indicating excellent potential to use the CNF reinforced starch foams as a potential structural insulation material. Future work will include investigating the effect of CNF reinforcement for higher total loadings and quantifying the hygroscopicity of the material and the effect of that on performance.

ACKNOWLEDGEMENT

The authors thank University of Maine Process Development Center for supplying cellulose nanofibrils in this study and thank Melanie Blumentritt's contribution of conducting the SEM images. This work partly supported by the USDA/NIFA under the Wood Utilization Research Program (Project 2010-34158-21182)

REFERENCES

1. Robert J. Moon; Ashlie Martini, John Nairn, John Simonsen and Jeff Youngblood. Chem. Soc. Rev. **2010**, 40, 3941-3944.
2. Ivo M.G. Martins; Sandra P. Magina, Lúcia Oliveira, Carmen S.R. Freire, Armando J.D. Silvestre, Carlos Pascoal Neto, Alessandro Gandini. Composite Science and Technology, **2009**, 69, 2163-2168
3. R. L. Shogren; J. W. Lawton, W. M. Doane and K. F. Tiefenbacher. Polymer, **1998**, 39, 6649-6655

4. Gregory M. Glenn; Syed H. Imam, and William J. Orts. MRS Bulletin, **2011**, 36, 696-702
5. Alain Dufresne and Michel R. Vignon. Macromolecules, **1998,** 31, 2963-2696
6. Gregory M. Glenn and Delilah W. Irving. Carbohydrates, **1995**, 72(2), 155-161
7. P. D. Tatarka, R. L. Cunningham. Journal of Applied Polymer Science, **1996**, 67, 1157-1176
8. L. Chen, S. H. Gordon, and S.H. Imam. Biomacromolecules, **2004**, 5, 238-244
9. Yogaraj Nabar; Jean Marie Raquez, Philippe Dubois, and Ramani Narayan. Biomacromolecules, **2005**, 6, 807-817
10. S. Bhatnagar and Milford A. Hanna. Cereal Chem. **1996**, 73(5), 601-604
11. Anna. J. Svagan, Lars A. Berglund, and Poul Jensen. Applied Materials & Interfaces, **2011**, 3, 1411-1417
12. R. L. Shogren; J. W. Lawton, K. F. Tiefenbacher. Industrial Crops and Products, **2002**, 16, 69-79
13. J. L. Willett, R. L. Shogren. Polymer, **2002**, 43, 5935-5947
14. S. Curvelo; A. J. F. de Carvalho, J. A. M. Agnelli. Carbohydrate Polymers, **2001,** 45, 183-188
15. Lorna J. Gibson and Michael F. Ashby, Cellular Solids; Structure & Properties, **1998,** p. 11
16. J. Shey, S.H. Imam; G.M. Glenn, W.J. Orts. Industrial crops and products, **2006,** 24, 34-40
17. Gypsum Assosiation. Gypsum board typical mechanical and physical properties (GA-235-10) **2010** www.gypsum.org
18. Xinying Wang; Jiming Zhong, Yimin Wang and Mingfang Yu. Carbon, **2006,** 44, 1560-1564
19. Zubaidah Mohammed Ali and Lorna J. Gibson. Soft Matter, **2013**, 9, 1580-1588
20. Houssine Sehaqui; Qi Zhou, Lars A. Berglund. Composites Science and Technology, **2011**, 71, 1593-1599
21. Chang-Qing Hong; Jie-Cai Han, Xing-Hong Zhang and Jian-Cong Du. Scripta Materialia, **2013**, 68, 599-602
22. T.M.I Mahlia; B.N. Taufiq, Ismail, H.H. Masjuki. Energy and Buildings, **2007**, 39, 182-187
23. Chuan-Liang Hsu and Dennis R. Heldman. International Journal of Food Science and Technology, **2004**, 39, 737-743
24. Gregory M. Glenn and Delilah W. Irving. Cereal Chem. **1995**, 72(2), 155-161
25. L. Petersson; I. Kvien, K. Oksman. Composites Science and Technology, **2007**, 67, 2535-2544

Mater. Res. Soc. Symp. Proc. Vol. 1621 © 2014 Materials Research Society
DOI: 10.1557/opl.2014.65

Degradation control of cellulose scaffold by Malaprade oxidation

Wichchulada Konkumnerd [1,2], Suong-Hyu Hyon[3], Kazuaki Matsumura[1]

[1]School of Materials Science, Japan Advanced Institute of Science and Technology, Japan

[2]Facalty of Science, Chulalongkorn University, Thailand

[3] Center for Fiber and Textile Science, Kyoto Institute of Technology, Japan

ABSTRACT

Study on oxidizing cellulose scaffold to dialdehyde cellulose by sodium periodate ($NaIO_4$) was carried out. Concentration of sodium periodate and the reaction time were effected for aldehyde introduction to cellulose scaffolds. Cellulose powder was dissolved in 1-butyl-3-methylimidazolium chloride, an ionic liquid, at 100°C and maintained at room temperature for 7 days, providing flexible cellulose scaffold. The cellulose scaffold was oxidized using periodate oxidation (Malaprade oxidation), which oxidizes carbohydrate by glycol cleavage to provide dialdehyde. Aldehyde groups introduced into cellulose were quantified by simple iodometry. Oxidized cellulose scaffold was degraded in the amino acid solution triggered by the reaction between aldehyde groups and amino groups. During immersion of the cellulose scaffolds in the amino acid solution, the mass loss of the scaffolds was evaluated by measuring of weight of oxidized cellulose scaffold before and after degradation.

INTRODUCTION

Tissue engineering is most usually assigned as the application of medical science and engineering principles in the design, erection, interpolation, growth and maintenance of living tissues. Normally, a three-dimensional structure, called scaffold, is necessary in tissue engineering applications since bioartificial tissues involve three-dimensional structures with cell multitude [1]. The evolution of the field of tissue engineering goes in parallel with the coherent demand for new scaffolding materials with definite properties such as controlled porosity and pore size distribution, biocompatibility and biodegradation. Among these materials, natural polymers predispose especially interest because of their biocompatibility, biodegradation, and exuberance. Polysaccharides are natural polymers that include cellulose, chitin, and starch. Cellulose is the most abundant polysaccharide found on earth, but cellulose has many limitations in utilization and manipulation such as its poor solubility in water or organic solvent, its solution is highly viscous and present gel formation ability. When a variety of cellulose derivative are produced, the oxidation is very important way [2,3]. Oxidation cellulose cause change in the structure and crystallinity of the results molecule, which affect its chemical and physical properties [4-6]. The oxidation can lead to a variety of polymer with difference functional group, so this can increase the range of cellulose application. We found that aldehyde introduced polysaccharide via Malaprade oxidation can be degraded at the glycoside bonds through the reaction with amino groups. In this study, we focus on the control of cellulose scaffolds degradation by oxidation and reaction with amine species.

EXPERIMENT DETIALS

Preparation of cellulose scaffold

Cellulose (0.323 g, 15% w/w for 1-butyl-3-methylimidazolium chloride (BMIMCl)) was dissolved in BMIMCl (2.15 g) by heating at 100 °C for 24 h and then adding sodium chloride for making porous [7]. After the solution was cooled to room temperature, it was sandwiched between silicon molds (diameter 10 mm, thickness 3 mm). The material was kept at room temperature for 7 days to form the gel material. The excluded BMIMCl was removed by washing with distilled water and freeze dried.

Synthesis of dialdehyde cellulose

Aldehyde cellulose scaffold was prepared by the oxidation of cellulose scaffold with sodium periodate [4]. Briefly, cellulose scaffold was immersed in 20 mL distilled water, and different amounts of sodium periodate (0.2-5.0 g). The reaction wasproceeded at 50°C for 1 h under gentle stirring. The oxidized cellulose scaffold was washed with distilled water 3 times and then freeze dried [7].

The aldehyde introduced into the cellulose scaffold were evaluated by simple iodometry. Briefly, 0.08 g oxidized cellulose scaffold was added to 20 mL of I_2 solution (0.05 mol/L), followed by addition of 20 mL of NaOH (1 mol/L). After the addition of 15 mL of H_2SO_4 (6.25 v/v%), the I_2 consumption by the reaction with aldehyde was titrated with 0.1 mol/L of $Na_2S_2O_3$ using the drop of aqueos 20 w/w% of starch solution as the indicator, where one mole of aldehyde group reacts with one mole of I_2 in alkaline condition, leading to the formation of carboxyl acid, and one mole of I_2 reacts with 2 moles of $S_2O_3^{2-}$ ion.

Degradation of oxidized cellulose scaffold *in vitro*

The oxidized cellulose scaffolds were degraded with glycine solution as amino acid solution. Briefly, 8 mg oxidized cellulose scaffold were immersed in glycine solution (5, 10, 15% w/v%). The weight of oxidized cellulose scaffolds were measured at various predetermined time.

Characterization of cellulose scaffold

The surface of the cellulose scaffold was characterized by scanning electron microscopy(SEM). For surface characterization, a dried scaffold was coated with palladium particles in a sputter coater and was observed at 100× magnification.

DISCUSSION

Aldehyde content of oxidized cellulose scaffold

Aldehyde cellulose scaffold were developed by using periodate oxidation (Scheme1,). The aldehyde groups were well introduced by periodated oxidation(Table 1).

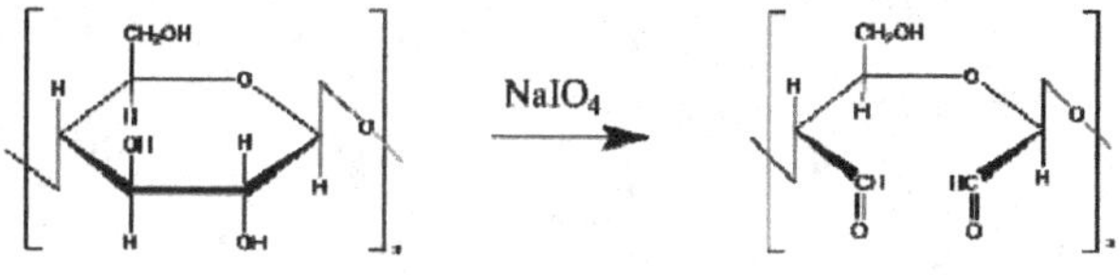

Cellulose   Oxidized cellulose

Scheme1. Aldehyde introduction by Malaprade oxidation of dextran

Table 1. Aldehyde content in oxidized cellulose scaffold.

% $NaIO_4$ introduced to cellulose scaffold	Mole of I_2	%Substitution of aldehyde
1	0.00005	10.18
5	0.00011	21.32
10	0.00016	32.70
15	0.00018	37.00
20	0.00021	42.58
25	0.00023	48.14

SEM

The scanning electron microscopy revealed porous architecture of scaffold (Fig. 1). Cellulose scaffold were fabricated by dissolved with ionic liquid and combined with NaCl leaching method [8]. Scaffold with NaCl (2b, 2c, 2d) had pore size larger than without NaCl (2a) when compare pore size which ratio of cellulose: NaCl, ratio 1:1(2c) had larger pores in size than those in ratio 6:4(2d). These results showed that the pore size of cellulose scaffold was controlled by varying the particle size of sodium chloride and ratio of cellulose per sodium chloride.

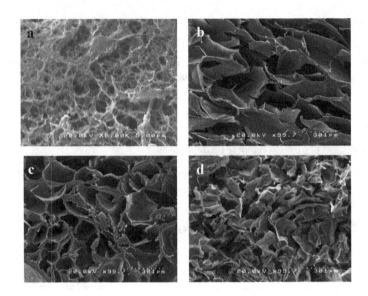

Figure 1. SEM micrograph of cellulose scaffold by ionic liquid without NaCl (2a), with NaCl 425 μm ratio cellulose: NaCl (1:1) (2b), with NaCl 250 μm ratio cellulose: NaCl (1:1) (2c) and with NaCl size 250 μm ratio cellulose: NaCl (6:4) (2b),

Studies on degradation *in vitro*

Degradation of oxidized cellulose scaffold with 0, 1, 5, 10, 15, 20, 25% $NaIO_4$ was investigated in 5% glycine solution for 16 hours (Table 2.). Oxidized cellulose scaffold with 5, 10, 15, 20, 25% $NaIO_4$ were completely degraded in 4, 2, 1, 0.75 and 0.5 hours respectively but 1% $NaIO_4$ oxidized cellulose scaffold was partially degraded and unoxidized cellulose scaffold was not degraded.

Weight loss of oxidized scaffolds were also investigated in 5% glycine solution for 16 hours (Fig. 2), weight loss tend to increase when time passed. The result well agreed with the result of Table 2. Oxidized cellulose scaffold oxidized with 5, 10, 15, 20, 25% $NaIO_4$ completely degraded in 4, 2, 1, 0.75 and 0.5 hours respectively, 1% $NaIO_4$ oxidized cellulose scaffold was partially degraded and unoxidized cellulose scaffold was not degraded. The degradation mechanisms are not revealed yet but the Maillard reaction might be related because of browning of the solution during degradation.

Table 2. Oxidized cellulose scaffold degradation in 16 hours

% oxidation NaIO$_4$	Time for completed degradation (hours)
0	No degradation
1	Partial degradation
5	4
10	2
15	1
20	0.75
25	0.50

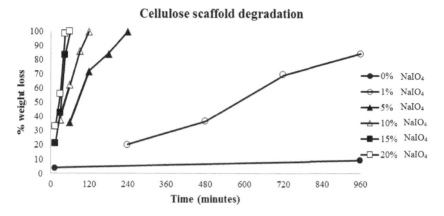

Figure 2. Weight loss of oxidized cellulose scaffold degradation in 16 hours.

CONCLUSIONS

The cellulose scaffold was fabricated by dissolving of cellulose powder in 1-butyl-3-methylimidazolium chloride as ionic liquid and control the pore size by adjust NaCl particles size and ratio of cellulose and NaCl. Aldehyde can be introduced to cellulose scaffold by control the concentration of sodium periodate (NaIO$_4$) and the degradation of oxidized cellulose scaffold were controlled in amino acid solution. From these results, we successfully developed the degradable cellulose scaffold by aldehyde introduction.

REFERENCES

1. H. Ko, C. Sfeir, and P.N. Kumta, *The Royal Society* **368**, 1981-1997(2010).
2. A. Elcin, *Cell Blood Sub* **34**, 407-418 (2006).
3. S. Meng, Y. Feng, Z. Liang, Q. Fu, and E. Zhang, *Transactions of Tianjin University* **11**, 250-254 (2004).
4. P. RoyChowdhury, and V. Kumar, *J Biomed Mater Res A* **76**, 300-309 (2005).
5. M. Singh, A. Ray, and P. Vasudevan, *Biomaterials* **3**, 16-20 (1982).
6. C. Tsioptsias, and C. Panayiotou, *Carbohyd Polym* **74**, 99-105 (2008).
7. B. Martina, K. Katerina, R. Miloslava, D. Jan, and M, Ruta, *Adv Polym Tech* **28**. 199-208 (2009).

Mater. Res. Soc. Symp. Proc. Vol. 1621 © 2014 Materials Research Society
DOI: 10.1557/opl.2014.66

Toward bioinspired nanostructures for selective vapor sensing: diverse vapor-induced spectral responses within iridescent scales of *Morpho* butterflies

Timothy A. Starkey[1], Peter Vukusic[1] and Radislav A. Potyrailo[2]*

[1] Natural Photonics Group, School of Physics, University of Exeter, EX4 4QL, UK.
[2] GE Global Research, 1 Research Circle, Niskayuna, NY 12309, USA.

* Corresponding author: Potyrailo@crd.ge.com

ABSTRACT

The iridescent colors of *Morpho* butterflies have captured scientific intrigue for over a century. However, only recently photonic structures of the wing scales of *Morpho* butterflies have inspired new ideas in the diverse areas of technology including sensing. In this study, we performed theoretical and experimental evaluation of vapor-induced reflectance changes of the *Morpho* scales. These experiments provided additional details of the origin and the magnitude of vapor response selectivity in these natural photonic nanostructures and facilitated our design and fabrication of highly selective biomimetic photonic nanostructures.

INTRODUCTION

The practical uses of chemical sensors are realized in a vast number of applications, such as in industrial processes, environmental studies, agriculture, clinical settings, and military technologies [1]. These applications stimulate growth in research into all aspects of high performance sensing. Whilst significant advances in receptor and transduction principles have been made, the ability to detect multiple volatiles and trace level molecules in real time remains challenging [2].

The biological world is known to contain a suite of ideas that stimulate technological applications, the most widely known commercial success is Velcro[TM] whose design was inspired by the barbs from the burdock plant [3]. Naturally evolved structures that manipulate light, which are commonly observed in animals [4]-[6] and plants [7], [8], are becoming increasingly studied to yield new ideas for technological exploitation.

Recently, the naturally formed photonic structures present in the wing scales of tropical *Morpho* butterflies have begun to provide novel bio-inspiration within the sensing community [9], [10]. The nanostructure present in scales that adorn the wings of iridescent *Morpho* butterflies comprises multi-layered thin film reflecting layers which interact with light through interference and diffraction phenomena [11] to reflect an often vivid blue color. Figure 1 shows the metallic reflection of two *Morpho* butterflies, and an example of the nanostructure responsible for this color appearance. In our recent study we have shown that the local chemical environment within the iridescent scales of the *Morpho* butterfly strongly influences the reflectivity upon low-concentration vapor exposures [12]. The combined role of this chemical environment, a gradient of surface polarity from the polar tops of ridges to the less polar bottoms, within the photonic structure of these scales allows the spectral transduction of volatile organic compounds (VOCs) due to their preferential adsorption [12]. These combined effects offer exciting prospects for the detection of VOCs within a single functionalized nanostructure rather than through traditional sensor array approaches.

Here we demonstrate that opportunities arise from utilizing *Morpho* nanostructures for sensing through experimental and theoretical results. First, we present the simulated reflectance of *Morpho* nanostructures upon uniform and preferential vapor adsorption. Then we experimentally demonstrate the selective vapor response of *Morpho* scales by measurement of reflectance at different angles.

Figure 1 - Photographs of (a) the blue *Morpho menelaus*, and (b) the pearly colored *Morpho sulkowskyi*. (c) A transmission electron micrograph (TEM) through the cross-section of a *M. sulkowskyi* scale revealing the reflecting lamella structure in the tree-like ridges. Their tops have a greater surface polarity then their bottoms. Scale bars: (a) 2.5 cm, (b) 4 cm, and (c) 500 nm.

THEORY

To theoretically study the optical effects that arise upon vapor adsorption onto the *Morpho's* iridescent scales, an optical model was developed. First, a rendered image of a single *Morpho* scale was created from high-resolution TEM images from cross sections through a *M. sulkowskyi* scale. This image was then input into commercially available finite element method (FEM) software (Ansoft HFSS) to numerically model the reflectance of the ridge structure over visible wavelengths. The ridge structure was assigned a complex refractive index of $n^* = 1.56 \pm 0.06i$, as previously reported in the literature [11]. Vapor adsorption was simulated as monolayer adsorption upon the ridge structure. Thicknesses of $t = 5$, 10 and 15 nm were selected based upon literature values [13] and values of refractive index $n = 1.3$, 1.4 and 1.5 were chosen. Total light reflectance of bare and vapor coated ridge structures was calculated for an infinite array of ridge structures, with 700 nm pitch (a single ridge is shown in Figure 2(a)), arising from a planar illumination with E-field oscillating parallel to the ridges.

1) Uniformly covered ridge structures

In order to explain the origin of the reported selective and sensitive vapor response [9] in *Morpho* scales, we first examined the reflectance from the *Morpho* structure upon uniform vapor adsorption upon the whole ridge (Figure 2(a)).

The calculated reflectance for uncoated structure and a coated structure with three thicknesses of refractive index $n = 1.5$ is shown in Figure 2(b). The measured reflectance from the wing of *M. sulkowskyi* is shown in Figure 2(c) as a qualitative comparison between the calculated and measured spectra. Upon vapor adsorption the reflectance red-shifts in wavelength. Figures 2(d), (e) and (f) demonstrate that this red-shift is dependent on the optical thickness ($n \times t$); the vapor layer of $t = 5$ nm with $n = 1.3$ shifts far less than the vapor layer with $t = 15$ nm with $n = 1.5$. We also note that there is little discernible difference between spectra of different vapor layers; the general shape is maintained for the optical thicknesses considered.

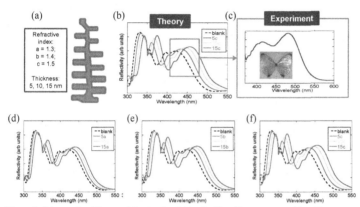

Figure 2 - (a) a rendered image of the simulated *Morpho*-like ridge structure with uniform vapor adsorption. The inset box provides a key for the different refractive indices and thicknesses of layers adsorbed in the simulated spectra displayed in panel (b) and panels (d)-(f). (b) comparison of modeled and (c) measured reflectance spectra, (dotted black line – represents the simulated reference spectra without vapor adsorption). (d), (e), and (f) show the reflectance of different thickness vapor layers for refractive indices $n = 1.3$, 1.4 and 1.5, respectively.

Experimental studies have shown that at high vapor concentrations different photonic materials, such as inverse opals or colloidal films, exhibit a linear relationship between the shift in peak wavelength and the vapor partial pressure [13]. In low-concentration regimes this wavelength-shift approach becomes almost undetectable and does not reveal information about the concentration of chemical species. Instead, a measurement of differential reflectance combined with multivariate analysis is a more common approach to access trends in the spectral data of such a measurement [14]. To compare these simulations to our earlier experimental results [9], these spectra are analyzed with Principal Components Analysis (PCA).

PCA is a robust and widely used analysis technique for extracting trends in high-order multivariate data sets [15]. The PCA technique reduces data dimensionality by projecting the original variables (wavelengths) onto a lower dimensional subspace [15]. The new variables are weighted sums of the original variables: these weighted sums are called Principal Components (PCs). The correlation coefficients between the original variables and the PC scores are called the loadings. These loadings indicate the contribution of each variable (wavelength) in the total variability of the PC. Data presented here has been analyzed using PLS_toolbox (Eigenvector Research, Inc.) operated with Matlab. In this work multivariate spectral analysis of the vapor response of *Morpho* scales was performed upon mean-centering [16] and autoscaling [17], [18] of simulated reflectance spectra and the measured differential reflectance spectra.

Preprocessing of the data was done by commonly employed mean-centering and autoscaling methods. Mean-centering method calculates the mean of each spectrum and subtracts this from the spectrum. Autoscaling method uses mean-centering followed by dividing each variable (wavelength) by the standard deviation of the spectrum [19]. Both methods were explored here in order to assess the importance of individual wavelengths in the differential reflectivity spectra.

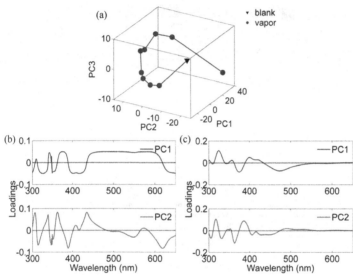

Figure 3 - PCA analysis of calculated reflectance of uniformly covered *Morpho*-ridge structures: (a) PCA scores plot (autoscaled) showing a non-selective response between model vapors. (b) and (c) display of the loadings of PC1 and PC2 for autoscaled and mean-centered processed data, respectively.

The results of PCA for the uniformly covered *Morpho* nanostructure are presented in Figure 3. Figure 3(a) shows a PC scores plot for the reference spectra and each vapor condition for the first 3 PCs. These simulated spectra form a track in the PCA space indicating that these vapor conditions do not produce the selective response reported previously in experiments. This trajectory follows the path of increasing optical thickness of adsorbed layer ($n \times t$). This result shows that the experimentally observed response does not arise solely from the coherent scattering associated with vapor uniformly coated upon the ridge structure. Although Figure 3(a) shows a PC scores plot of autoscaled spectra, the PC scores plot of mean-centered spectra looked similar.

The loadings for PC1 and PC2 are shown in Figures 3(b) and 3(c) for both preprocessing methods. The percentage of variance captured in the first, second and third Principal Components for autoscaled spectra was 69.3%, 25.2% and 4.7%, respectively, and for mean centered spectra this was 84.3%, 13.6% and 1.73%, respectively. The loadings for the autoscaled preprocessed model, shown in Figure 3(b) indicate that the wavelengths between ~325 to ~650 nm contribute to the variance in the autoscaled data set. The loadings for the mean centered preprocessed model, shown in Figure 3(c) indicate that the wavelengths between ~325 to ~500 nm contribute most to the variance in the mean-centered data set. As expected [19] the loadings from the autoscaled and mean-centered data sets were different.

2) Gradient covered ridge structures

In the pursuit of understanding how selectivity is achieved within a single nanostructure, our recent study employed a suite of experimental techniques to study the chemical properties of the iridescent scales [12]. These techniques included TEM characterization, optical microscopy characterization, chemical composition analysis, and vapor exposures analysis. Results from these methods revealed a local chemical environment within the ridges, specifically, a gradient of chemical polarity [12].

To explore this theoretically, models with three spatial regions of vapor adsorption were considered. The hypothesis of preferential vapor adsorption was simplified to regions of vapor adsorption upon the top, middle and bottom of the ridge (see Figure 4(a)). The reflectance spectra simulated for these model vapor adsorptions are shown in Figure 4(b). Visually these spectra display a great deal more diversity than those of uniform vapor adsorption. As noted previously, the vapor layers cause the spectra to red-shift, as expected from standard multilayer interference [20]. In addition to a wavelength red-shifting, different vapor locations result in discernable alterations in reflectance. First, we notice that vapor on the bottom of the ridge results in smaller spectral shifts than vapor adsorbed on the top. Second, we observe that the spectral shapes are noticeably different for vapor in different spatial regions. Both of these observations arise due to the changes in coherent scattering conditions associated with the spatial position of preferential vapor adsorption; the larger relative change in lamellar layer thickness by the addition of vapor to the top of the ridge structure explains its relative larger visual spectral diversity.

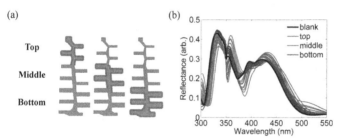

Figure 4 - (a) schematic representation of modeled gradient-type vapor adsorption on the top, middle and bottom portions of ridge structure, and (b) reflectance spectra for vapor regions depicted in (a), for refractive indices n = 1.3, 1.4 and 1.5 with layer thicknesses of t = 5, 10 and 15 nm.

The results of PCA for the gradient-type vapor covered *Morpho* nanostructure simulations are presented in Figure 5. Figure 5(a) shows a PC scores plot for the reference spectra and each vapor condition for the first 3 PCs. In contrast to the PCA scores plot for uniform vapor adsorption (Figure 3(a)), the projection of points for each spatial location considered forms three tracks. Both preprocessing methods reveal three tracks in PCA space. Each track originates from the reference point and extends, as before, to points of increasing adsorbed optical thickness. Each track arises from each spatial region within the ridge, that is to say PCA analysis has differentiated the different spatial regions associated with the differences observed in the spectral domain.

Therefore, by using this simplistic consideration of preferential vapor adsorption FEM simulation results, we are able to reproduce the experimentally observed selective-type vapor response. This provides further evidence that the preferential adsorption due to local chemical environments coupled within the interaction of light with these hierarchical ridge structures allows for the selective transduction of a chemical signal through reflectance measurements.

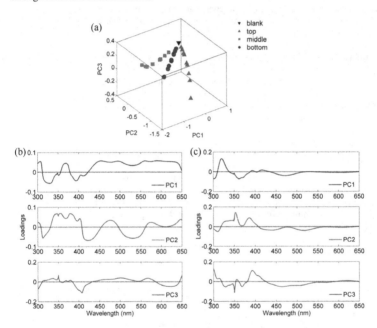

Figure 5 - PCA analysis of the calculated reflectance from *Morpho*-ridge structures with an assumed gradient of vapor adsorption: (a) PCA scores plot (mean-centered) showing a selective response that arises between model vapors coated to the top, middle and bottom spatial regions. (b) and (c) display the loadings of PC1, PC2 and PC3 for autoscaled and mean-centered processed data, respectively.

The loadings for PC1, PC2 and PC3 are shown in Figures 5(b) and 5(c) for both preprocessing methods. The percentage of variance captured in PC1, PC2 and PC3 for autoscaled spectra was 49.0%, 35.6% and 10.0%, respectively, and for mean centered spectra this was 62.8%, 30.3% and 3.7%, respectively. If we compare these figures to those of the PCA analysis for the uniformly coated structure we see that the more variance is captured with higher PCs. This greater spread in our PCA space indicates that the gradient type adsorption has larger spectral diversity than uniform adsorption.

EXPERIMENT

In previous studies of vapor-induced changes in reflectance from the iridescent scales, a bifurcated probe arrangement was used to illuminate and then measure reflectance at the same angle of incidence [9], [12], we shall call this a back-scatter measurement. To date, this approach has allowed the measurement of selective and sensitive responses. To further explore a different measurement of reflectance, and the resulting vapor responses, we present a comparison between this 'back-scatter' measurement, and a measurement of the specular reflection, that we shall term the 'forward-scatter'.

To measure this, a bifurcated probe was used to illuminate (using a halogen light source), and to record (using a fiber-optic spectrograph) the reflectance from the blue iridescent scale of $M.$ $rhetenor$. The illumination/ back-scatter-reflectance probe was positioned at $\sim 30°$ from the normal to the sample surface. A ~ 2 mm diameter beam spot illuminated the scale. For the measurement of forward-scattered light, another bifurcated probe was positioned at $\sim -30°$ from the normal. The schematic in Figure 6 (a) depicts the orientation of both measurement probes with respect to the orientation of the ridges in the butterfly's scale.

To perform vapor exposures, a bubbler system was used. Generated vapors were flowed with a 450-mL/min total flow rate from a 1/8-inch-diameter tube onto the butterfly wing at an angle of approximately 45 degrees to the surface. The vapor flow was orientated to flow along the ridges of the photonic structure. Vapors were flowed with 0.04, 0.07 and 0.10 P/P_0 concentrations, where P is the partial vapor pressure, and P_0 is the saturated vapor pressure. The change in reflectance, ΔR, upon vapor exposure was measured as ΔR (%) = $(R/R_0 - 1) \times 100\%$.

The comparison of back- and forward-scatter measurements is shown in Figure 6(b), for water, propanol and methanol at a concentration of 0.07 P/P_0. In both measurements the differential reflectance has very similar magnitudes and spectral shapes. The coherent scatter of light from the hierarchical $Morpho$ structures are known to produce broad blue hemispherical scattering to create a conspicuous appearance [11], [21], so this is perhaps not surprising.

The temporal profiles for vapor exposures for the back- and forward-scatter measurements are shown in Figures 6(c) and 6(d), respectively. Once again the general trends remain the same, however, differences in the dynamic evaporation for methanol vapor cycles are observed. In the back-scatter measurement (Figure 6(c)), the 525 nm trace begins to smoothly return to the baseline reflectance at the end of the vapor cycle, whilst the forward-scatter measurement displays a rapid return to the baseline before the ΔR signal increases again, Figure 6(e) shows two fixed wavelengths for both measurements for this methanol cycle. Such observations can reveal information about the interactions of vapors within the structure.

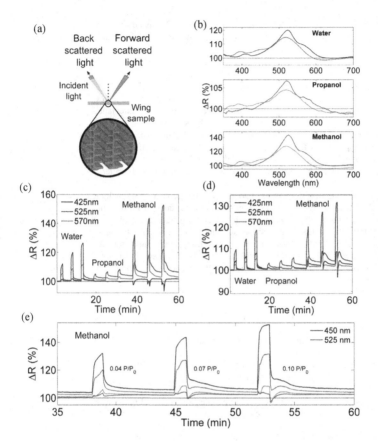

Figure 6 - (a) diagram illustrating the experimental arrangement of illumination of the sample, and collection of scattered light, with respect to the orientation of the ridges. (b) back-scattered (solid lines) and forward-scattered (dotted lines) measured spectra for water, propanol and methanol vapors at 0.07 P/P$_0$ concentration. (c) and (d) show the temporal response of back- and forward-scattered light, respectively for 1-min vapor exposures of 0.04, 0.07 and 0.10 P/P$_0$ concentrations at 3 wavelengths. (e) provides a comparison of back- (solid lines) and forward- (dotted line) scatter reflection at two wavelengths during the methanol exposures.

PCA scores plots of the back- and forward-scatter measurements are shown in Figures 7(a) and 7(b), respectively. In both Figures, water and propanol vapor results have two tracks from the reference point; the methanol exposures do not display this behavior. This effect as present because there was not sufficient time to allow the signal to return to the baseline reading before the first methanol exposure. Thus a binary mixture of propanol and methanol is present on the ridge structure, this point is circled in Figure 7. The captured variance (autoscaled) of PC1 and PC2 for the back-scatter measurement are 72.7% and 22.0%, respectively, and 76.2% and 24.2% for the forward-scatter measurement. The interesting result here is that the scores plots give similar qualitative results, with comparable variance captured on each PC, for both measurements. Such apparent insensitivity in measurement conditions may have some practical application in remote measurements where the fine control of the optical measurement position is not possible.

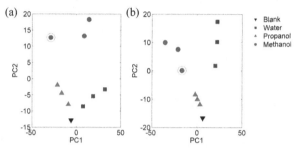

Figure 7 - (a) and (b) PCA scores plots (autoscaled) for back- and forward-scattered reflectance respectively, for water, propanol and methanol at 0.04, 0.07 and 0.10 P/P$_0$. The circled point represents the 0.04 P/P$_0$ vapor exposure.

CONCLUSIONS

Here we have discussed the vapor-induced spectral responses within the scales of *Morpho* butterflies through theoretical optical simulations and experimental vapor-exposure measurements. First, simulated reflectance spectra were coupled with their PCA analysis and were used to examine the effects of uniform and preferential vapor adsorption on the ridge structure. The modeled preferential vapor adsorption explained our earlier reported experimental selective vapor responses. Analysis of the PCA loadings highlighted the wavelengths that contributed to the variation within each PC. Second, wing scales of *M. rhetenor* were exposed to different vapors and the change in reflection for back- and forward-scattered reflected light was studied in detail. Temporal profiles show slightly different dynamics between measurements, whilst PCA analysis of both, back-scatter and forward-scatter, measurements provides similar response selectivity.

At present, we are completing vapor-response characterization of biomimetic photonic structures that we have fabricated using diverse nanofabrication technologies. Our multivariable sensor concept initially explored using natural photonic structures of *Morpho* butterflies has proved to be extremely useful for the development of these biomimetic sensors. Results of these studies will be reported shortly.

ACKNOWLEDGMENTS

The authors acknowledge the financial support from DARPA Contract W911NF-10-C-0069 and in part from GE's Advanced Technology research funds.

REFERENCES

[1]   G. Korotcenkov, *Chemical Sensors: Comprehensive Sensor Technologies. Volume 1: General Approaches*. New York: Momemtum Press, 2009.

[2]   J. P. Fitch, E. Raber, and D. R. Imbro, "Technology challenges in responding to biological or chemical attacks in the civilian sector.," *Science (80-.).*, vol. 302, no. 5649, pp. 1350–4, Nov. 2003.

[3]   P. Forbes, *The Gecko's Foot: How Scientists are Taking a Leaf from Nature's Book*, 1st ed. Harper Perennial, 2006, p. 356.

[4]   P. Vukusic and J. R. Sambles, "Photonic structures in biology," *Nature*, vol. 424, pp. 852–855, 2003.

[5]   S. Kinoshita and S. Yoshioka, *Structural Colors in Biological Systems: Principles and Applications*, 1st ed. Osaka University Press, 2005.

[6]   T. Starkey and P. Vukusic, "Light manipulation principles in biological photonic systems," *Nanophotonics*, vol. 2, no. 4, pp. 289–307, 2013.

[7]   D. W. Lee, "Iridescent blue plants," *Am. Sci.*, vol. 85, no. 1, pp. 56–63, 1997.

[8]   S. Vignolini, E. Moyroud, B. J. Glover, and U. Steiner, "Analysing photonic structures in plants.," *J. R. Soc. Interface*, vol. 10, no. 87, p. 20130394, Oct. 2013.

[9]   R. A. Potyrailo, H. Ghiradella, A. Vertiatchikh, K. Dovidenko, J. R. Cournoyer, and E. Olson, "Morpho butterfly wing scales demonstrate highly selective vapour response," *Nat. Photonics*, vol. 1, no. 2, pp. 123–128, Feb. 2007.

[10] A. D. Pris, Y. Utturkar, C. Surman, W. G. Morris, A. Vert, A. D. Zalyubovskiy, T. Deng, H. T. Ghiradella, and R. A. Potyrailo, "Towards high-speed imaging of infrared photons with bio-inspired nanoarchitectures," *Nat. Photonics*, vol. 6, no. February, pp. 195–200, 2012.

[11] P. Vukusic, J. R. Sambles, C. R. Lawrence, and R. J. Wootton, "Quantified interference and diffraction in single Morpho butterfly scales," *Proc. R. Soc. B Biol. Sci.*, vol. 266, no. 1427, pp. 1403–1411, Jul. 1999.

[12] R. A. Potyrailo, T. A. Starkey, P. Vukusic, H. Ghiradella, M. Vasudev, T. Bunning, R. R. Naik, Z. Tang, M. Larsen, T. Deng, S. Zhong, M. Palacios, J. C. Grande, G. Zorn, G. Goddard, and S. Zalubovsky, "Discovery of the surface polarity gradient on iridescent Morpho butterfly scales reveals a mechanism of their selective vapor response.," *Proc. Natl. Acad. Sci. U. S. A.*, vol. 110, no. 39, pp. 15567–15572, Sep. 2013.

[13] H. Yang, P. Jiang, and B. Jiang, "Vapor detection enabled by self-assembled colloidal photonic crystals.," *J. Colloid Interface Sci.*, vol. 370, no. 1, pp. 11–8, Mar. 2012.

[14] R. A. Potyrailo, Z. Ding, M. D. Butts, S. E. Genovese, and T. Deng, "Selective Chemical Sensing Using Structurally Colored Core-Shell Colloidal Crystal Films," *IEEE Sens. J.*, vol. 8, no. 6, pp. 815–822, Jun. 2008.

[15] P. C. Jurs, G. a Bakken, and H. E. McClelland, "Computational methods for the analysis of chemical sensor array data from volatile analytes.," *Chem. Rev.*, vol. 100, no. 7, pp. 2649–78, Jul. 2000.

[16] M. Bahram, "Mean centering of ratio spectra as a new method for determination of rate constants of consecutive reactions.," *Anal. Chim. Acta*, vol. 603, no. 1, pp. 13–9, Nov. 2007.

[17] J. H. Perkins, E. J. Hasenoehrl, and P. R. Griffiths, "The use of principal component analysis for the structural interpretation of mid-infrared spectra," *Chemom. Intell. Lab. Syst.*, vol. 15, no. 1, pp. 75–86, May 1992.

[18] M. Sanchez, E. Bertran, and L. Sarabia, "Quality control decisions with near infrared data," *Chemom. Intell. Lab. Syst.*, vol. 53, pp. 69–80, 2000.

[19] B. M. Wise, N. B. Gallagher, J. M. Shaver, W. Windig, R. S. Koch, and P. L. S. T. Version, *PLS_Toolbox 4.0: for use with MATLAB*. Eigenvector Research Inc, 2006.

[20] M. F. Land, "The physics and biology of animal reflectors.," *Prog. Biophys. Mol. Biol.*, vol. 24, pp. 75–106, Jan. 1972.

[21] S. Yoshioka and S. Kinoshita, "Wavelength–selective and anisotropic light–diffusing scale on the wing of the Morpho butterfly," *Proc. R. Soc. B Biol. Sci.*, vol. 271, pp. 581–587, 2004.

Mater. Res. Soc. Symp. Proc. Vol. 1621 © 2014 Materials Research Society
DOI: 10.1557/opl.2014.2

Integrative Chemistry toward Biosourced SiC Macrocellular Foams Bearing Unprecented Heat Transport Properties

Simona Ungureanu[1,2], Marc Birot[2], Gérard Vignoles[3], Christophe Lorette[4], Gilles Sigaud[1], Hervé Deleuze[2], Rénal Backov[1]

[1]Centre de Recherche Paul Pascal, UPR 8641-CNRS, Université de Bordeaux,
115 Avenue Albert Schweitzer, 33600 Pessac, France
[2] Université de Bordeaux, Institut des Sciences Moléculaires (ISM) UMR 5255 CNRS,
351 Cours de la Libération, 33405 Talence, France
[3] Université de Bordeaux, Laboratoire des Composites Thermostructuraux, UMR 5801 CNRS-UB1-CEA-Snecma Propulsion Solide,
3 Allée de la Boétie, 33600 Pessac, France
[4] CEA, Laboratoire des Composites Thermostructuraux, UMR 5801 CNRS-UB1-CEA-Snecma Propulsion Solide,
3 Allée de la Boétie, 33600 Pessac, France

ABSTRACT

Black liquor is a by-product of the paper mill Kraft process that deserves more valorization than its present use as low-grade fuel. In this work, SiC/C composite foams were prepared for the first time from concentrated emulsions by carbothermal reduction of bio-sourced precursors combining sodium silicate by lignin at 1400°C. The composition of the materials was determined by XRD, FTIR and Raman analyses. Their porous structure was characterized by SEM, mercury intrusion porosimetry, and nitrogen sorption, while their thermal properties were measured by TGA and dynamic DSC. Concerning their heat transport properties, we found out that when the starting lignin content was increased, the final C/Si ratio, the specific surface area and the heat diffusivity increased as well. Its high values were attributed to a cooperative effect between radiative heat transfer and the presence of partially graphitized carbon.

INTRODUCTION

Trends toward transformation of waste/biomass to valuable materials are growing stronger because of the depletion of natural resources and increasing of greenhouse emission [1]. Moreover, chemistry of materials relies strongly on rational design over all length scales, where final enhanced functionality will ensure the overall synthetic pathway to be applied. From this way of thinking has recently emerged the concept of *Integrative Chemistry* [2]. One synthetic path combines sol-gel chemistry with lyotropic mesophases and concentrated direct emulsions to promote either inorganic Si(HIPE) [3] or hybrid organic-inorganic foams named Organo-Si(HIPE) [4] (the acronym HIPE relies for High Internal Phase Emulsion) [5]. Recently, we used Si(HIPE) macrocellular foams as hard template to produce the parent carbonaceous macrocellular foams [6] bearing standard application in Li-ion battery electrodes and chemical electro-capacitors devices, hydrogen storage when modified with Li(BH₄) [7]. This process has been also extended to the morphosyntheses of boronitrite BN(HIPE) [8] and silicon carbide/carbon composite SiC/C(HIPE) [9]. Indeed, silicon carbide (SiC) is an important non-oxide ceramic associated to a set of unique properties such as high thermal stability, high heat conductance addressed through small heat expansion, endurance toward high temperature oxidation and corrosion, overall chemical inertness and bearing semi-conductor behavior at high temperature [10]. There are mainly two crystallographic forms of SiC: the high temperature (> 1500 °C) hexagonal α-phase that presents two commercially available polytypes, namely 4H- and 6H-SiC, and the low temperature (< 1500 °C) cubic β-phase [11].

In this study we have synthesized for the first time biosourced SiC/C composites macrocellular foams using renewable and inexpensive sources, namely Kraft lignin for carbon, and sodium silicate (Na_2SiO_3) for silicon. The synthesis relies on a one pot route where macroporosity is promoted by emulsifying castor oil within the continuous hydrophilic lignin phase. In a second section, thermophysical properties of these composite macrocellular foams are provided in details.

EXPERIMENTAL DETAILS

*Materials.*Black liquor (BL) (45% dry matter) came from Smurfit Kappa Kraft paper mill (Facture plant, France) and was used as received. Sodium silicate powder (Na2SiO3), Cremophor EL and castor oil were purchased from Aldrich. Epichlorohydrin (EP) was purchased from Acros Organics. All these chemicals were used as received.

HIPE preparation. In typical experiments, 20 g of black liquor (45% dry matter, carbon content C=40% wt) and sodium silicate in molar ratios C/Si = 30, 20, 15 were mixed for 30 min. This mixture, along with 1 g of Cremophor EL and epichlorohydrin (2.2 g, 2.3 mmol), were placed in one of the chamber of the emulsification device prior to the addition of castor oil (20 g or 16 g). Emulsification was performed using a homemade system already described [12].The viscous, black emulsion obtained was placed in tightly closed PTFE cylindrical moulds and polymerized for 24 h at 60 °C in an oven. The resulting polyHIPEs monoliths were washed by refluxing successively with ethanol (72 h) in a Soxhlet extractor and dried in a vacuum oven at room temperature to constant weight. The monoliths were then heat-treated twice under argon atmosphere. In the first treatment, the acquired monoliths were heated at 1000 °C for 2 h with a first plateau at 300 °C for 5 h (heating rate of 2 °C min⁻¹) and a second at 700 °C (heating rate of 0.5 °C min⁻¹) in order to decompose the organic species. In the second treatment, the temperature was raised to the final value for carbidization: 1400 °C for 3 h with a heating rate of 2 °C min⁻¹. Final SiC/C macrocellular foams were labeled hereafter as Bio-SiC/C(HIPE)x, x being the starting C/Si molar ratio (30, 20, or 15).

Characterizations. The macroscopic void space morphology of the final foams was observed by scanning electron microscopy (SEM) in a Hitachi TM-1000 microscope. High-resolution SEM observations were performed with a Jeol 6700 FX microscope (accelerating voltage of 10 kV). Intrusion/extrusion mercury measurements were performed using a Micromeritics Autopore IV 9500 analyzer. The specific surface area was determined by N_2 adsorption measurements performed on a Micromeritics ASAP 2010 apparatus. XRD experiments were carried out on a PANalytical X'pert MPD diffractometer with Bragg-Brentano θ-θ geometry, CuKα1 radiation (40 kV, 40 mA, λCu= 1.5418 Å). Thermogravimetric analyses (TGA) were carried out under an air flow (5 cm3 min⁻¹) using a heating rate of 5 °C min⁻¹. The apparatus was a Netzsch STA 409 instrument. Raman spectra were recorded at 297 K on a Xplora confocal micro-Raman spectrometer (Horiba Jobin Yvon), in backscattering geometry at 2.33 eV laser energy (532 nm) with a typical spectral resolution of 1.7 cm-1 (objective 50 ×). Fourier-transformed infrared (FTIR) spectra were obtained using a Nicolet 750 FTIR spectrometer using ground samples that were mixed with KBr. Thermal analyses to obtain the foam calorific capacities (Cp) were conducted using the modulated temperature mode of a TA-Q2000 DSC apparatus. The measurements of the heat capacities of the three samples were performed on a TA Q2000 apparatus. Thermal diffusivities were estimated using a heating laser source (Coherent Sapphire™ 488-20, 20 mW @ 488 nm) and an IR camera (Cedip FLIR SC 7000). The sample surface was subjected to a laser step impulse, and the front surface temperature was recorded as a function of time. The frame rate was 760 per second, and the pixel size was 0.142 mm.

DISCUSSION

Industrial sources of lignin come from the pulp and paper industry, mainly from the Kraft process that represents the major part of the industrial plants presently in use in the world [13]. The major part of the degraded lignin biopolymer thus solubilized contributes to the dark brown pollution load of the so-called black liquors [14]. This very important amount of biomass waste, more than 10^7 tons per year in the European Union alone, is now merely used as in-house low-grade fuel [15]. In this study, along with lignin as carbon precursor we employed sodium silicate powder as silicon source to synthesize SiC/C composites. Sodium silicate cumulates three advantages: firstly, the powder can fully mix with the black liquor carbon precursor, which is an alkaline aqueous solution. After complete dissolution of Na_2SiO_3, the intimate mixing of the reactants can be performed. Secondly, owing to this silicon source, we can remarkably reduce the cost of the produced composites since the price of sodium silicate is much cheaper than silica sol or tetraethylorthosilicate (TEOS). Thirdly, sodium silicate is an environment-friendly precursor. In this context, two concentrated emulsions were prepared at 55 vol % in oil as described in experimental details leading to materials labeled Bio-SiC/C(HIPE)$_{30}$ and Bio-SiC/C(HIPE)$_{20}$ (molar ratio of C/Si of 30, 20) and one emulsion at 40 vol % in oil leading to material labeled Bio-SiC/C(HIPE)$_{15}$ for a molar ratio C/Si=15. When the amount of sodium silicate was high, the emulsion was not stable anymore, and a reverse emulsion through a phase inversion phenomenon was obtained. Once the emulsions are formulated, the organic polymerization is promoted through the adding of epichlorohydrin.

Figure 1. (a) Typical Bio-SiC/C(HIPE)$_{20}$ monolith after thermal treatment at 1400°C, (b) SEM micrograph and (inset) high-resolution SEM micrograph of the Bio-SiC/C(HIPE)$_{20}$, (c) Mercury intrusion porosimetry for Bio-SiC/C(HIPE)$_{20}$.

Figure 1a shows the appearance of the one of the composites after the carbothermal reduction. Despite the pyrolysis reaction, and associated 60% of overall weight loss, these monolith-type materials (Figure 1a), exhibit an interconnected porosity (Figure 1b). High-resolution SEM observations (Figure 1b inset) show the presence of nanowires that compose the foams walls. This feature has been reported already [16]. The macroscopic pore size distribution obtained from mercury intrusion porosimetry is polydisperse as shown by Figure 1c, with an important population of diameters at 1 µm, a porosity of 80-85% and bulk densities between 0.11-0.25 $m^2 \cdot g^{-1}$. Considering the carbon material nitrogen sorption isotherms (Figure 2a), it depicts a type I-IV, characteristic of micro/mesoporous materials. This porosity is likely textural, promoted by the surface of the nanowires observed in Figure 1b inset. The mesopore surface area of the foams is almost constant and fall around 180 $m^2 \cdot g^{-1}$, suggesting that there is no big difference from one foam wall texture to another. Also, the specific surface area (BET) are increasing when increasing the amount of carbon in the starting emulsions and are around to 570, 429 and 356 $m^2 \cdot g^{-1}$ for C/Si molar ratio of 30, 20, and 15 respectively. XRD patterns of the samples produced at 1400 °C are presented in Figure 2b. The position of the diffraction peaks clearly indicates the formation of cubic β-SiC. The SiC synthesis proceeding through the carbothermal reduction reaction is therefore complete, since no crystalline SiO_2 phase was detected. A very weak peak visible in the diffraction pattern at ~ 26.1 ° indicates the presence of graphitic carbon in the samples and can be explained by remaining carbon in the final composites. The fact that no silica is present at the end of the carbidization step is also proven by infrared spectroscopy investigations (results not presented here). In order to check if the lignin was partially graphitized (not detectable by XRD), Raman spectroscopy was used to investigate this point. Two bands appear around 1590 and 1350 cm^{-1} (Figure 3a) that correspond to the graphite sp_2 (E_{2g}) carbon band (usually labeled G band) and to the defects band (usually labeled D band), respectively. As reported by Tuinstra and Koenig [17] the ratio of the carbon band intensities, $I_{D\ band}/I_{G\ band}$, depicts a constant graphitized crystallite coherence length, L_a, close to 2 nm for the materials Bio-SiC/C(HIPE)$_{15}$ and Bio-SiC/C(HIPE)$_{30}$, and 7nm for the Bio-SiC/C(HIPE)$_{20}$. Thermogravimetric analyses in air have been performed to assess the amount of carbonaceous matter associated to these SiC/C composite foams (Figure 2c). As expected, the weight decrease around 650 °C comes from the burn off of the carbon, whereas the weight increase derives from the oxidation of SiC, and certainly some SiOC, into SiO_2 [18].

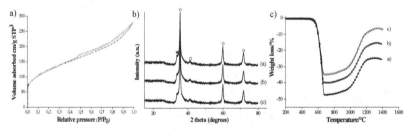

Figure 2 (a) Nitrogen adsorption/desorption curves of Bio-SiC/C(HIPE)$_{20}$, (b) XRD patterns of C/SiC composites at different C/Si molar ratios: a) Bio-SiC/C(HIPE)$_{30}$ b) Bio-SiC/C(HIPE)$_{20}$ c) Bio-SiC/C(HIPE)$_{15}$. The diffraction peak assigned to α-SiC is indicated by an * and those assigned to β-SiC by opened circles, (c) TG curves in air of different SiC/C composites: a) Bio-SiC/C(HIPE)$_{30}$ b) Bio-SiC/C(HIPE)$_{20}$ c) Bio-SiC/C(HIPE)$_{15}$.

The C/Si molar ratios calculated from the TG curves are 3, 2.2 and 1.7 for the composite foams Bio-SiC/C(HIPE)$_{30}$, Bio SiC/C(HIPE)$_{20}$ and Bio-SiC/C(HIPE)$_{15}$ respectively. We have also studied the foams thermal behavior. We first have conducted some dynamic DSC thermal analyses in order to obtain the heat capacity C_p vs. composition and temperature. Three different scanning rates of heating were used to check the influence of this parameter on the measurement of the heat capacity on such macroporous and highly divided compounds. Indeed, the apparent specific heat capacity of the Bio-SiC/C(HIPE)$_x$ appears dependent on the heating rate, especially at temperatures above 400 K, in contrast with the heat capacities measured on the sapphire reference (Figure 3b). Also, the heat capacities at 298 K are rather close, between 0.65 for Bio-SiC/C(HIPE)$_{30,\ 15}$ and 0.70 J·g^{-1}·K^{-1} Bio-SiC/C(HIPE)$_{20}$. Heat diffusivity has been obtained using an analytical approximation of the laser spot heating problem [9]. Neglecting convective heat losses, the temperature field at the surface evolves with time according to the following formula [19]:

$$\theta(r,t) = \frac{q}{4\pi\lambda r}\,\mathrm{erfc}\!\left(\frac{r}{\sqrt{4\alpha t}}\right) \quad (1)$$

where $\theta(r,t) = T(r,t) - T_0$ is the temperature in excess above the ambient (K), q is the heating flux (W·m^{-2}), λ is the heat diffusivity (W·m^{-1}·K^{-1}), $\alpha = \lambda/(\rho C_p)$ is the heat diffusivity (m^2·s^{-1}) and erfc is the complementary error function. In the limit of r^2/t tending to large values, the erfc term tends to 1, allowing identification of the thermal conductivity by fitting $q/4\pi\theta$ versus r. Nonetheless, plotting $r\theta(r,t)$ versus $r/\sqrt{t}$ and fitting by an erfc function yields extremely correct results for the foams, as illustrated in Figure 3c. The results at 25 °C are collected in Table 1. The high values of heat diffusivity probably rest partly on heat transfer by radiation. Moreover, the diffusivity increase with the carbon content and can be assigned to a cooperative effect between

Table 1. Materials heat transport properties.

Material	Heat capacity $C_{p,app}$ J.g^{-1}.K^{-1}	Thermal diffusivity α mm^{2}s^{-1}	Heat conductivity λ (W.m^{-1}.K^{-1})
Bio-SiC/C(HIPE)$_{30}$	0.65 ± 0.03	108 ± 5	17.6 ± 1.6
Bio-SiC/C(HIPE)$_{20}$	0.70 ± 0.02	63 ± 2	7.5 ± 0.5
Bio-SiC/C(HIPE)$_{15}$	0.65 ± 0.02	52 ± 2	3.7 ± 0.3

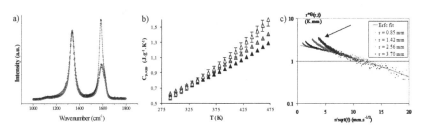

Figure 3 (a) Raman spectra of the SiC/C composites obtained at 1400 °C under argon (o: Bio-SiC/C(HIPE)$_{15}$; +: Bio-SiC/C(HIPE)$_{20}$; ◊: Bio-SiC/C(HIPE)$_{30}$); (b) Specific heat capacity as a function of temperature for Bio-SiC/C(HIPE)$_{30}$; (c) Example of transformed time data for several points lying apart from the spot, and complementary error function curve fitting for Bio-SiC/C(HIPE)$_{30}$

an increasing radiative heat transfer on one hand, and the amount of graphite crystallites that should increase with the starting amount of carbon, on the other hand.

CONCLUSION

Concentrated emulsions of castor oil in Kraft lignin/sodium silicate aqueous solutions could be polymerized with epichlorohydrin. The resulting monoliths were converted to biosourced SiC/C composite foams by carbothermal reduction of the silicate by lignin at 1400 °C. XRD, Raman and FTIR spectroscopies revealed that β-SiC and partially graphitized carbon were formed, and that no silica was detected at the end of the carbidization step. The graphite nucleation and growth might be initiated by the SiC itself. SEM observations and porosity measurements confirmed that porosity was hierarchical: the macroporosity being induced by the emulsion, while the meso- and microporosities were promoted by the nanowires surface. Different experiments were carried out by varying the lignin/silicate ratio. When the starting lignin content was increased, the final C/Si ratio, the specific surface area and the heat diffusivity increased as well. Its high values were attributed to radiative heat transfer and to the presence of partially graphitized carbon.

REFERENCES

1. G. W. Huber, S. Iborra, A. Corma, *Chem. Rev.* **106**, 4044 (2006).
2. Brun, N.; Ungureanu, S.; Deleuze, H.; Backov, R. *Chem. Soc. Rev.* **40**, 771 (2011).
3. Carn, F.; Colin, A.; Achard, M.F.; Deleuze, H.; Sellier, E.; Birot, M.; Backov R. *J. Mater. Chem.* **14**, 1370 (2004).
4. Ungureanu, S.; Birot, M.; Laurent, G.; Deleuze, H.; Babot, O.; Julián-López, B; Achard, M.F.; Popa, M.I.; Sanchez, C; Backov, R. *Chem. Mater.* **19**, 5786 (2007).
5. Barby, D.; Haq, Z. *Eur. Pat. 0,060,138 (to Unilever)* (1982).
6. Brun, N.; Prabaharan, S.; Morcrette, M.; Sanchez, C.; Pécastaing, G.; Derré, A.; Soum, A.; Deleuze, H.; Birot, M.; Backov, R. *Adv. Funct. Mater.* **19**, 3136 (2009).
7. Brun, N.; Janot, R.; Sanchez, C.; Deleuze, H.; Gervais, C.; Morcrette, M.; Backov, R.; *EnergyEnviron. Sci.* **3**, 824 (2010).
8. Alauzun, J. G.; Ungureanu, S.; Brun, N.; Bernard, S.; Miele, P.; Backov R. Sanchez, C. *J. Mater. Chem.* **21**, 14025 (2011).
9. Ungureanu, S.; Birot, M.; Lorrette, C.; Vignoles, G.; Derré, A.; Babot, O.; Deleuze, H.; Soum, A.; Pécastaing, G.; Backov, R. *J. Mater. Chem.* **21**, 14732 (2011).
10. Klemm, H.; Herrmann, M.; Schubert, C. *J. Eng. Gas Turbines Power* **122**, 13 (2000).
11. Shaffer, P.T.B. *Acta Cryst.* **B25**, 477 (1969).
12. Lépine, O.; Birot, M.; Deleuze, H. *Coll. Polym. Sci.* **286**, 1273 (2008).
13. Smook, G.A. *Handbook of pulp and paper technologists.* Vancouver: Angus Wilde Publications 3rd edition (2002).
14. Cardoso, M.; Domingos de Oliveira, E.; Passos, M.L. *Fuel* **88**, 756 (2009).
15. Monte, M.C.; Fuente, E.; Blanco, A.; Negro, C. *Waste Man.* **29**, 293 (2009).
16. Shivani, B.M.; Ajay, K.M.; Rui, W.K.; Bhekie, B.M. *J. Am. Ceram. Soc.* **92**, 3052 (2009).
17. Tuinstra, F.; Koenig, J.L. *J. Chem. Phys.* **53**, 1126 (1970).
18. Hasegawa, G.; Kanamori, K.; Nakanishi, K.; Hanada, T. *Chem. Mater.* **22**, 2541 (2012).
19. Carslaw, H.S.; Jaeger, J.C. *Conduction of Heat in Solids*, Oxford: Oxford University Press 2nd ed., p. 261 (1959).

Mater. Res. Soc. Symp. Proc. Vol. 1621 © 2014 Materials Research Society
DOI: 10.1557/opl.2014.120

Preparation of Porous B-type Carbonate Apatite with Different Carbonate Contents for an Artificial Bone Substitute

Toshimitsu Tanaka[1], Tomohiko Yoshioka[1], Toshiyuki Ikoma[1] and Junzo Tanaka[1]
[1]Department of Metallurgy and Ceramics Science, Tokyo Institute of Technology, 2-12-1 Ookayama, Meguro-ku, Tokyo 152-8550 JAPAN

ABSTRACT

B-type carbonate apatite (B-Cap) powders were prepared by a wet method using $Ca(OH)_2$ suspension and H_3PO_4 solution including $NaHCO_3$ as a carbonate source, and porous B-CAp ceramics with two different amounts of carbonate contents were fabricated by sintering freeze-dried mixtures of the powders and gelatin composite. The porous B-CAp ceramics prepared had three-dimensionally interconnected pores. The sinterability of B-CAp ceramics was dependent on the chemical composition, especially sodium content and vacancy of OH site, and the carbonate contents did not directly influence the dissolution rate of porous B-CAp ceramics.

INTRODUCTION

Porous hydroxyapatite (HAp) ceramics have been widely used as an artificial bone in clinical treatments because HAp has high osteoconductivity and the porous structure enable the cells to penetrate inside the ceramics where new bone tissues are regenerated. However, in some cases, porous HAp ceramics unfortunately caused bone fractures due to its low bioabsorbability. Therefore, there is a strong demand to improve bioabsorbability of HAp material.

Carbonate apatite (CAp) in which carbonate ions are substituted for the sites at phosphate groups (B-type; B-CAP), hydroxyl groups (A-type) or both groups (AB-type) in HAp, has higher dissolution rate compared with HAp [1]. The dissolution amount of non-sintered CAp powders has been described to be linearly proportional to increasing carbonate content [2,-3]. On the other hand, porous sintered B-CAp ceramics with low carbonate contents dissolved rapidly compared to those with high carbonate contents in simulated physiological solution (SPS) at 37°C and pH 7.3, containing sodium (137 mM), chloride (177 mM), and HEPES buffer (50 mM) [4]. Therefore, there is controversy regarding the influence of carbonate content on the dissolution behavior of porous B-CAp ceramics.

In this study, porous ceramics of sintered B-CAp with different carbonate contents were prepared by sintering the composite of B-CAp/gelatin mixture. The effect of carbonate contents on the sinterability and the dissolution property in the acetate buffer at pH 5.5 was investigated.

EXPERIMENT

Preparation of porous B-CAp ceramics with different carbonate content

B-CAp powders were prepared by a wet chemical method, in which H_3PO_4 solution and $Ca(OH)_2$ suspension including different amounts of $NaHCO_3$ as a carbonate source were used. A reaction formula is shown as follows;

$$(10-x)Ca(OH)_2 + (6-x)H_3PO_4 + xNaHCO_3$$
$$\rightarrow Ca_{(10-x)}Na_x(PO_4)_{(6-x)}(CO_3)_x(OH)_2 + (18-2x)H_2O$$

where x was set to be 1 and 3 in the experiment. The abbreviation of B-CAp1 and B-CAp2 means x= 1 and 3 in the reaction formula.

Firstly, the appropriate amounts of $NaHCO_3$ were added into 0.5M $Ca(OH)_2$ suspension under stirring. 0.3M of H_3PO_4 solution was dropped into the suspension including the appropriate amounts of $NaHCO_3$ at the dropping speed of 40ml/ min. The B-CAp powders were obtained after separation of solid and liquid with centrifuging and freeze-dried. The B-CAp powders were mixed with ultrapure water at weight ratio of 3:10 and the suspensions were then well stirred using ultra-sonication. The gelatin powder (type-A, Sigma) was slowly added into the suspensions and stirred at 50°C to be the weight ratio of B-CAp and gelatin of 4:1 for 2 h. The mixture at 50°C was poured into a cylindrical mold at 8 mm in diameter, frozen at -20°C for 24 h and then freeze-dried. The composite obtained was sintered at 700°C for 2 h at the heating rate of 150°C/h.

Characterization Methods

The crystalline phase of the porous B-CAp ceramics was identified by using X-ray diffractiometry (XRD; X'Pert-MPD, PANAlytical). KBr powder was mixed into the crushed ceramics to produce disk pellets for Fourier transform infrared spectroscopy (FT-IR). The substitution sites of carbonate ion in the apatite lattice were analyzed by FT-IR (FT-IR4100N, JASCO) with a diffuse reflectance method in the range of 4000-400 cm^{-1} at the resolution of 4 cm^{-1}. Carbonate contents in the porous ceramics were measured by TG-DTA (Thermo plus EVO/TG-DTA, Rigaku) using alumina sample as reference. To obtain the TGA curve, 10 mg of the crushed ceramic was placed in a platinum pan and heated at a rate of 20 °C/min. The elemental analysis of Ca, P, and Na ions for each ceramics was conducted by an inductively coupled plasma optical emission spectrometry (ICP-OES; Prodigy ICP, Leeman Labs, Inc. U.S.A). The ceramics were dried at 110°C to constant weight and then dissolved in 5% HNO3 solution. The surface morphology was observed with a scanning electron microscope (SEM; JSM 6510LV, JEOL) after being coated with platinum at 40mA for 40s. In the dissolution tests, each ceramic at around 30 mg was immersed in acetate buffer at 37°C for 180 min under stirring at 300 rpm. The concentration of acetate buffer was 0.08 M and the pH was 5.50 ± 0.03. The Ca concentration was measured with a calcium ion meter (F- 55, HORIBA). To adjust the slight difference of the mass of the specimens, the measured value of Ca ions released in the acetate buffer from the ceramics was divided by the total Ca mass in the weight of immersed ceramics.

RESULT and DISCUSSION

Physical and Chemical Characterization

XRD patterns for porous B-CAp ceramics as shown in Fig.1 indicated that all samples had the diffractions identified as hexagonal apatite except the diffraction detected at 29.0° in B-CAp1. Watanobe et al. [5] described that the diffraction was divided into two belonged to B-CAp after heating at higher temperature. Therefore, both diffraction at 28.8 and 29.0° could be assigned to the Miller index of 210, which means the diffraction at 29.0° is not assigned to second phase. Each ceramic sintered at 700°C is a single phase of apatite.

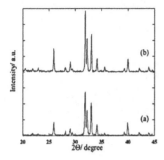

Fig.1 XRD patterns of porous B-CAp ceramics of (a) B-CAp1 and (b) B-CAp2

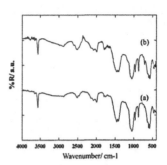

Fig.2 FT-IR spectra of porous B-CAp ceramics of (a) B-CAp1 and (b) B-CAp2

The FT-IR spectra of the ceramics are shown in Fig. 2. In all spectra, there are the three phosphate bands; v_1 vibration at 959cm^{-1}, v_3 vibration at 1054 cm^{-1}, v_4 vibration at 570 and 603 cm^{-1}, and two carbonate bands; v_2 vibration at 874 cm^{-1}, v_3 vibration at 1410 and 1457 cm^{-1}, which are characteristic of B-CAp [6]. In only B-CAp2, the peak at 714 cm^{-1} assigned to v_4 vibration of carbonate ions was detected [7]. Feki et al. described that the v_4 vibration of carbonate ions was detected in B-CAp powder with 10.1~22.4 wt% of carbonate ions [8]. Therefore, we assumed that B-CAp2 contained over 10wt% carbonate.

Table 1 shows carbonate contents and chemical compositions of each ceramic. The electrical neutrality of the compositions was considered. V_{Ca} and V_{OH} represent the vacancy of Ca and OH sites, respectively. The carbonate contents of B-CAp1 and B-CAp2 were 5.31 and 10.12%, respectively. In terms of the contents of sodium and the vacancy of OH, those of B-CAp2 were higher than those of B-CAp1. The amounts of the vacancy of Ca sites in each ceramics were almost same.

Table 1. Carbonate contents and chemical compositions of each ceramic

	Carbonate content(%)	Chemical composition
B-CAp1	5.31	$[Ca_{8.61}Na_{0.46}V_{Ca0.93}][(PO_4)_{4.86}(CO_3)_{1.14}][(OH)_{0.82}V_{OH1.18}]$
B-CAp2	10.12	$[Ca_{7.39}Na_{1.69}V_{Ca0.92}][(PO_4)_{4.06}(CO_3)_{1.94}][(OH)_{0.41}V_{OH1.59}]$
		[Ca site] [PO$_4$ site] [OH site]

Porous structure Characterization

     Fig. 3 shows the SEM images of each sample. The pore sizes of B-CAp1 and B-CAp2 were 272±65 and 252±77 µm (n=10), respectively. It is well known that a freeze-dried gelatin has hornets' nest like porous structure. SEM images exhibited that the porous CAp ceramics fabricated had three-dimensional pores based on the gelatin structure. In addition, the ceramics had interconnected pores which should enable cells to penetrate inside the ceramics.

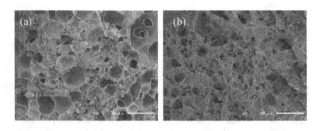

Fig.3 SEM images of porous B-CAp ceramics of (a) B-CAp1 and (b) B-CAp2

Sinterability and compressive strength

     Fig. 4 shows that the values of linear shrinkage and grain size of each ceramics. The linear shrinkage of B-CAp1 and B-CAp2 were 27.8 ± 1.2 and 44.1 ± 0.53 % (n=3), while the grain size of B-CAp1 and B-CAp2 were 0.21 ± 0.05 and 0.25 ± 0.06 µm (n=30), respectively. Fig. 5 shows the compressive strength and porosity of each ceramics. The compressive strength of B-CAp1 and B- CAp2 were 2.27±0.06 and 6.07±0.43 MPa (n=3), while the porosities of B-CAp1 and B- CAp2 were 77.3±0.73 and 59.3±2.9 % (n=3), respectively.

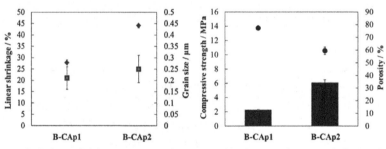

Fig.4 Linear shrinkage (♦) and grain size(■) of CAp1 and CAp2

Fig.5 Compressive strength (bar) and porosity (●) of CAp1 and CAp2

Lafon et al. reported that B-CAp with the chemical formula $Ca_{9.05}V_{Ca0.95}$ $(PO_4)_{5.05}(CO_3)_{0.95}$ $(OH)_{1.05}V_{OH\,0.95}$ did not shrink at all at 700°C [9]. In contrast, Na bearing B-CAps in this experiment shrank at 700°C. The difference indicates that sodium contents and vacancies of OH can improve the sinterability of B-CAp ceramics drastically because the sodium contents and vacancies of OH in B-CAp1 and B-CAp2 were larger than that in B-CAp Lafon et al. prepared. The reason why the shrinkage of B-CAp2 was larger than that of B-CAp1 can be explained using this tendency due to larger sodium contents and vacancy of OH.

Rodriguez-Lorenzo et al. described that the decrease in porosity was related to the increase in density and lower porosity led to larger compressive strength [10]. Based on the fact that the increases in shrinkage and in density occur at the same time, the increase in shrinkage is thought to be proportional to the decrease in porosity. From these results, larger sodium contents and vacancies of OH could improve the sinterability of B-CAp ceramics.

Dissolution behavior of each sample

Fig. 6 shows the dissolution rate of the samples as a function of soaking time. At 180 min, the dissolution rate was larger on B-CAp1 than on B-CAp2. To consider the influence of porous structure, we divided the dissolution rate by surface area because porosity of each sample influenced the BET surface area. The BET surface area of B-CAp1 and 2 were 12.5 and 15.9 m^2/g. However, this calculation did not influence the order of the dissolution rate. This indicates that porous structure never influences the dissolution rate. The results implied that porous B-CAp ceramics with the low carbonate content dissolve rapidly compared with that with the high carbonate content, which was not in agreement with literatures [2, -3]. However, Habibovic et al. reported that porous B-CAp ceramics with 3wt% carbonate had higher dissolution rate than that with 5 wt% in SPS at 37°C and pH 7.3 [4]. Therefore, we concluded that the contents of carbonate in the apatite lattice did not well correlate with dissolution rate of CAp.

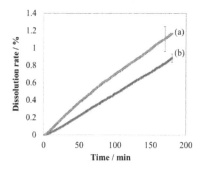

Fig.6 Dissolution rate of porous B-CAp ceramics of (a) B-CAp1 and (b) B-CAp2

CONCLUSIONS

In this study, we succeeded in preparing porous B-CAp ceramics by sintering the freeze-dried mixtures of the powders and gelatin composite at 700°C. In the observation of the sample surface, all samples had the macro and interconnected pores. The pore sizes of B-CAp1 and B-CAp2 were large enough for cells to penetrate inside the ceramics. The compressive strength of CAp1 and CAp2 were 2.27±0.06 and 6.07±0.43MPa. The sinterability of B-CAp was dependent on the chemical composition, especially sodium content and vacancies of OH site, which decreased the porosity and increased the compressive strength. The dissolution rate at 180 min was larger on B-CAp1 than on B-CAp2, and this order didn't change after considering the BET surface area. From these results, we concluded that the contents of carbonate ions in the apatite lattice and the porosity did not well correlate with dissolution rate of carbonate apatite.

ACKNOWLEDGEMENTS

The authors thank Mr. Kuwayama and Mr. Higaki of Kuraray Co., Ltd. for their fruitful discussion.

REFERENCE

1. M. Hasegawa, *J Bone Joint Surg [Br]* 85-B (2003) 142-147
2. Y. Doi et al, *J Biomed Mater Res* **39** (1998) 603-610.
3. LeGeros R.Z et al. *Caries Res* 17 (1983) 419–429
4. P. Habibovic et al. *Acta Biomaterialia* 6 (2010) 2219–2226
5. N. Watanobe et al. *MRS Proceedings* Volume 1301 (2011) 39-44
6. I. R. Gibson et al. *J Biomed Mater Res* 59 (2001) 697-708
7. S. Markovic et al. *Biomed. Mater* 6 (2011) 045005
8. H. El Feki et al. *Calcif Tissue Int* 49 (1991) 269-274
9. J. P. Lafon et al. *J. Eur. Ceram. Soc.*, 28 (2008) 139-147
10. Rouriguez-Lorenzo et al. *biomater* 22 (2001) 583-588

Mater. Res. Soc. Symp. Proc. Vol. 1621 © 2014 Materials Research Society
DOI: 10.1557/opl.2014.3

Science of Swimming and the Swimming of the Soft Shelled Turtle

Shinichiro Ito[1]

[1] Department of Mechanical Engineering, Kogakuin University, 1-24-2 Nishi-Shinjuku,
Shinjuku-ku, Tokyo, 1680065, Japan

ABSTRACT

Swimming is dynamically a part of the hydrodynamic field and can be considered as a
field of the optimal control motion. Animals move by instinct according to the situation which
they are confronting with. Therefore, their instinctive motion is optimal most of the time. The
movement of animals can be classified roughly into two kinds: the fast motion with the
maximum speed and the motion with the minimum energy consumption. Considering the foreleg
of the soft shelled turtle as a flat plane, several sets of movement of the foreleg were observed
and calculated theoretically. The theoretical results agreed the observation results in the both
cases with the maximum speed and the minimum energy consumption. Applying the theoretical
movement of the soft shelled turtle foreleg to human movement in swimming, the general S-
shaped pull stroke is the minimum energy consumption motion in free-style. It became clear that
there was a different stroke for generating the maximum speed in free-style. That was the soft
shelled turtle style of fast swimming, the I-shaped pull strokes. In 2002 when the author
announced this theory, there was only one fast swimmer whose free-style swimming strokes
coincidentally accorded with the I-shaped pull with fewer numbers of strokes at that time. He
was the Olympic gold medalist Ian Thorp. Now the I-shaped stoke has become main stream in
free style.

INTRODUCTION

Advancing movement of an animal in water can be roughly divided into two categories,
locomotion of the maximal efficiency (the minimal energy consumption mode) for a usual
motion and that of the maximal speed (the maximal thrust mode) for an urgent evacuation or a
predatory action instinctively.
James Counsilman [1] is one of the first to apply physical principles to try to understand
the mechanism of propulsion. His study, where underwater cameras were used for the first time,
showed a skillful swimmer moved his arms in an S-shaped pattern over the body axis in rolling
motion. This arm motion produced lift like a propeller on an airplane. He suggested that
propeller-like diagonal sculling motion, S-shaped pull stroke, was used by skilled swimmers,
acknowledging the importance of lift forces.
A speed is determined at the steady state of the body where the resistance of the whole
body and propelling force are balanced. In order to reduce resistance of the body, a shark skin
swimsuit has been developed. On the other hand, what affects the propulsion for its increase? In
free style swimming, thrust force is mainly generated by movements of the arms (Hollander et al.
[2]). They reported that the propelling force ratio of arms to legs was from 10:1 to 6:1. Moreover,

the lift-drag characteristics of an arm are similar to those of a palm, says Berger et al [3]. Therefore, the lift-drag force characteristics of a palm can be considered as a main factor that rules over the impelling force

Azuma and the author [4] studied optimal ways of paddling locomotion theoretically and verified by observing swimming locomotion of reeve's and soft-shelled turtles in a water channel. While the equations obtained by the above are applied to swimming forms of humans based on their instinct, the author introduces freestyle swimming strokes for the maximal efficiency and for the maximal speed.

Human's instinctive motion of the maximal speed might have been altered by intelligence in the swimming history. For competitive swimming, an operation of the maximal propelling force is desirable. In the freestyle swimming, forms of the operation were calculated by using equations of turtles' instinctive locomotion [5].

THEORY

In general, when an object moves through a fluid, force R acting upon it can be decomposed into two components: a drag force D acting opposite to a direction of an advancing direction and a lift force L acting perpendicular to it.

An inclined paddle/hand by a tilt angle θ is moved diagonally with a driving velocity U and a driving angle δ while the body moves with an advancing velocity V. As a result, a relative velocity W with an angle of attack α to the hand was shown in Figure 1. The drag force D acts opposite to the direction of W and the lift force L acts perpendicular to W. A thrust force T is the component of the resultant force R to the advancing direction. As an aspect ratio changes, significant differences in characteristics of lift-drag forces appear with varying sweepback angles ψ whose convention is shown in Figure 2. Sweepback angles of a palm, defined by

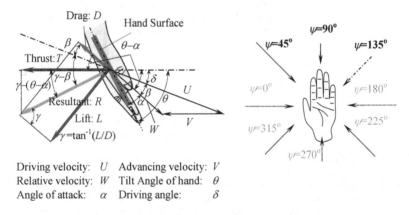

Driving velocity: U Advancing velocity: V
Relative velocity: W Tilt Angle of hand: θ
Angle of attack: α Driving angle: δ

Figure 1. Forces acting on a hand paddle

Figure 2. Sweepback angle ψ defined by Schleihauf [6]

Schleihauf [6], ψ=135°, 90° and 45° correspond to the catch, the pull and the finish phase on freestyle stroke shown in figure 1 respectively. In a constant swimming speed or in a quasi-steady state, propelling force T is equivalent to the dynamic whole body derivative resistance D_{DP}. Relation of each variable above are formulated and solved simultaneously by using lift-drag characteristics of the paddle/palm.

Thrust force T, power P and efficiency η could be defined with them:

$$T = \frac{1}{2}\rho W^2 SC_R \cos(\gamma + \alpha - \theta) = \frac{1}{2}\rho W^2 SC_T \tag{1}$$

$$P = \frac{1}{2}\rho W^2 USC_R \cos(\gamma - \beta) = \frac{1}{2}\rho W^3 SC_P \tag{2}$$

$$\eta = TV / P \tag{3}$$
$$= \cos(\gamma + \alpha - \theta) / \{(U/V)\cos(\gamma - \beta)\}$$

where

$$L = \frac{1}{2}\rho W^2 SC_L, \quad D = \frac{1}{2}\rho W^2 SC_D \tag{4a, b}$$

$$C_R = \sqrt{C_L^2 + C_D^2} \tag{4c}$$

$$\gamma = \tan^{-1}(L/D) \tag{5}$$

In the above variables, the following geometrical relations are established:

$$W/V = (U/V)\cos\beta - \sqrt{1 - (U/V)^2 \sin^2\beta} \tag{6}$$

$$\tan(\theta - \alpha)$$
$$= (U/V)\sin\beta / \{(U/V)\cos\beta - (W/V)\} \tag{7}$$

$$\delta = \theta - (\alpha + \beta) \tag{8}$$

$$\beta = \sin^{-1}\{\sin(\theta - \alpha)/(U/V)\} \tag{9}$$

EXPERIMENT

The experimental objects are a soft-shelled turtle, *Pelodiscus sinensis* and a reeve's turtle, *Chinemys reevesii*. In order to investigate the paddling locomotion in a steady swimming condition, a turtle was left in a stream condition. For their maximum speed test, the water speed was controlled voluntarily for the position of the swimming turtles being stayed at a part of test section shown in figure 3. Also let the turtles swim freely in a still water to capture the minimum energy consumption motion. The swimming forms were filmed from both the left and the bottom sides simultaneously.

Also a plaster replica of a hand of an excellent swimmer was made and wind tunnel tests were performed with it to measure its aerodynamic performance. As an aspect ratio changes, significant differences in characteristics of lift-drag forces appear with varying sweepback angles ψ whose convention is shown in figure 2. Sweepback angles of the hand, ψ=135°, 90° and 45° correspond to the catch, the pull and the finish phase in freestyle stroke respectively. With the variation of sweepback angle ψ, the hydrodynamics characteristics against angle of attack were

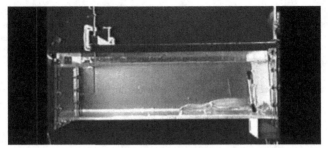

Figure 3. An observation circulation water tank for a swimming

measured in a Reynolds number (9.7×10^5 as representative length of the hand) by an wind tunnel (flow velocity 30 m/s) almost equivalent to the relative inflow velocity (2 m/s) against the water to the hand of actual swimming.

RESULTS AND DISCUSSION

By the observation of the turtle's swimming stroke, the tilt angle of fore and hind paddles indicated almost 90° to the axis of locomotion in the power stroke of maximum speed mode, while the paddles were inclined backwards in the power stroke of minimum energy consumption mode, as shown in figures 4 respectively. The theoretical results agreed the observation results in the both cases with the maximum speed and the minimum energy consumption. Applying the theoretical movement of the soft shelled turtle foreleg to human movement in swimming with the original data of lift and drag coefficient of a hand palm replica measured by the wind tunnel experiment, the maximum speed mode of the soft shelled turtle in free-style swimming was lead as the I-shaped pull stroke shown in figure 7. The general S-shaped pull stroke was lead as the minimum energy consumption motion in free-style as indicated in figure 8. To explain in detail,

Figure 4(a). Maximum speed mode   **Figure 4(b).** Minimum energy
consumption mode

Figures 4. Tilt angle difference of respective paddles in the power stroke of turtles

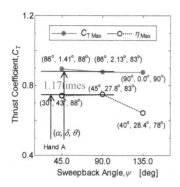

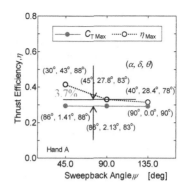

Figure 5. Difference between thrust coefficient, C_T of the maximum thrust and that of the maximum efficiency η among different sweepback angles ψ

Figure 6. Difference between thrust efficiency, η of the maximum thrust and that of the maximum thrust coefficient, C_T among sweepback angle ψ

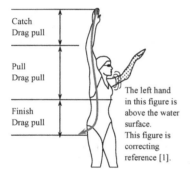

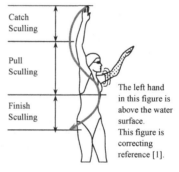

Figure 7. The stroke phases of I-Shaped pull

Figure 8. The stroke phases of S-Shaped pull

(i) Figure 5 shows the differences in thrust coefficient C_T between the swimming of the maximal efficiency and that of the maximal thrust. The highest thrust coefficient $C_T = 0.878$ is original data of lift and drag coefficient of a hand palm replica measured by the wind tunnel experiment, the maximum speed mode of the soft shelled turtle in free-style swimming was lead as the I-shaped pull stroke shown in figure 7. The general S-shaped pull stroke was lead as the minimum energy consumption motion in free-style as indicated in figure 8.

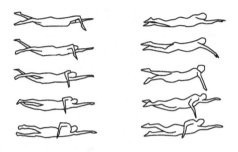

Figure 9(a). I-Shaped pull   **Figure 9(b).** S-Shaped pull

Figures 9. Decomposed images of I-Shaped pull (drag used) stroke and S-Shaped pull (lift used) stroke form in free-style swimming

(ii) The difference in thrust efficiency between the both methods is demonstrated in figure 6. It shows a remarkable result that the difference of thrust efficiency in the maximum thrust, η_{TMax}=32.9%, and the maximum efficiency, η_{Max} =29.2%, is only 3.7%. The maximum efficiency η_{max} can be obtained when θ=90° or θ=85°. The angle of attack α at each of the maximum points is the stage where lift-to-drag ratio is the maximum, γ_{max}= [tan^{-1}(L/D)]$_{max}$. It corresponds to the S-shaped pull motion shown in figure 8. It accords with underwater motion of skilled swimmers.

Achievements

It was shown above that drag type swimming is the fastest swimming form which generates the maximum thrust. There existed an actual swimmer with I-shaped pull had 4 individual world records at the time the author presented this theory. His name was Ian Thorpe, an Australian swimmer. Figures 9 shows underwater images compared with I-shaped and S-shaped stroke in side-view. As for S-shaped pull swimming shown in figure 9(a), the swimmer's elbow is extended in the catch phase. On the other hand, in I-shaped pull in figure 9(b), the elbow is bent immediately after putting into water.

This I -shaped pull stroke in free-style has become a de facto standard in a competitive swimming and drag pull stroke is considered to be applicable to all 4 styles of swimming. The improvement of swimming technique has revived to be an important mean to update records after the swimsuit regulation revision.

CONCLUSIONS

In conclusions, the propelling force can be divided into two optimal kinds, the mode of the maximum thrust force and that of the minimum energy consumption by the lift-drag curve of a hand.

By the calculations based on the hydrodynamic characteristics of hand, the S-shaped pull stroke, a popular form for the conventional free-style, is resulted as a form of the maximum efficiency mode utilizing lift and drag force by hands.

Concerning the mode of the maximum propelling force which should be used in a competitive swimming, the hand should be driven along the body axis to the advancing direction for drag forces to be used entirely like I-Shaped pull stroke.

REFERENCES

1. Counsilman, J.E., *Science of Swimming,* Prentice-Hall, Englewood Cliffs, N.J (1968).
2. Hollander, A. P. et al., Contribution of the legs to propulsion in front crawl swimming, *Swimming Science V*, pp.17-29. Human Kinetics Publishers, Champaign, IL (1987).
3. Berger, M.A.M. et al., Hydrodynamic drag and lift forces on human hand/arm models, *J. Biomechanics*, 28, pp.125-133(1995).
4. Ito, S and A. Azuma, Mechanical Analysis of Paddling Concerning in Turtles Locomotion, *Proc. 1ˢᵗ Int'l Symposium on Aqua Bio-Mechanisms*, 30-36 (2000).
5. Ito, S. and Okuno, K., A Fluid Dynamical Consideration for Arm stroke in Swimming, *Biomechanics and Medicine in Swimming IX*, pp.39-44, Pub. de l'univ. de Saint-Etienne (2003).
6. Schleihauf, R.E., A hydrodynamic analysis of swimming propulsion, *Swimming III*, pp70-109. University Park Press, Baltimore, MD (1979).

Mater. Res. Soc. Symp. Proc. Vol. 1621 © 2014 Materials Research Society
DOI: 10.1557/opl.2014.174

Effect of Microtubules Hierarchy on Photoinduced Hydrogen Generation and Application to Artificial Photosynthesis

Kosuke Okeyoshi[1,2], Kawamura Ryuzo[1], Ryo Yoshida[2], and Yoshihito Osada[1]

[1] RIKEN Advanced Science Institute, 2-1 Hirosawa Wako-shi, Saitama 351-0198, Japan
[2] Department of Materials Engineering, Graduate School of Engineering, The University of Tokyo, 7-3-1 Hongo, Bunkyo-ku, Tokyo, 113-8656, Japan

ABSTRACT

Several strategies have been explored from viewpoint of biomimetics to accomplish artificial photosynthesis by using macromolecules as a medium such as liposomes, supramolecules, and hydrogels.[1] Differing from disordered solution systems in which multiple components such as photosensitizer and catalytic nanoparticle are diffusively mixed, the photochemical reactions occur efficiently in medium due to maintenance of the dipersibility of the components and specific molecular arrangement. Here we attempt to clarify the effect of medium hierarchy for photoinduced electronic transmission among multiple components. By conjugating each component on tubulin and integrating them via self-assembly to microtubules, ideal component arrangements with optimum distance for the electronic transmission will be possible.

INTRODUCTION

We designed and fabricated $Ru(bpy)_3^{2+}$-conjugated microtubules for photoinduced H_2 generation. By conjugating $Ru(bpy)_3^{2+}$ on microtubules, the structure enables to avoid aggregation among the $Ru(bpy)_3^{2+}$ that cause loss of photoenergy in self-quenching process. The hierarchy of tubulin/microtubules is controlled by using 6-thioguanosine-5'-triphosphate (6-thio-GTP) which maintains the structure as tubulin, or guanosine-5'-(β,γ-methylene)triphosphate (GpCpp) which maintains the structure as microtubules (Fig. 1). When the conjugated $Ru(bpy)_3^{2+}$ is excited by photoenergy, it will effectively give electron to Pt nanoparticle and H_2 generates.

EXPERIMENTAL

Preparation of Tubulin. To remove microtubules-associated proteins, tubulin was purified from porcine brain using a high concentrations of PIPES buffer (1 M PIPES, 20 mM ethyleneglycol-bis(β-aminoethyl ether)-N,N,N',N'-tetraacetic acid (EGTA), 10 mM $MgCl_2$; pH adjusted to 6.8 using HCl).[2] The tubulin concentration was determined by measuring the absorbance at 280 nm using an extinction coefficient of 115,000.

$Ru(bpy)_3^{2+}$ Conjugation and Stoichiometric Estimation. $Ru(bpy)_3^{2+}$-conjugated tubulin was prepared using Bis(2,2'-bipyridine)-4'-methyl-4-carboxybipyridine-ruthenium N-succinimidyl ester-bis(hexafluorophosphate) by amine coupling according to previous studies.[3,4] The

Ru(bpy)$_3$$^{2+}$-tubulin stoichiometry, i.e., Ru(bpy)$_3$$^{2+}$ molecules/tubulin was estimated to be 3.1 ± 0.1/2.0 on average. The efficiency of labeling by the amine coupling was 31%. The concentration of Ru(bpy)$_3$$^{2+}$ was determined by measuring the peak absorbance at 460 nm. The conjugation of Ru(bpy)$_3$$^{2+}$ on tubulin was confirmed by observing the loaded gel on an optical band-pass filter (λ = 450 ± 25 nm, BPB 45; Fujifilm). Tubulin concentration was determined by the contrast of the stained band for SDS-PAGE, comparing with that of a non-conjugated tubulin suspension.

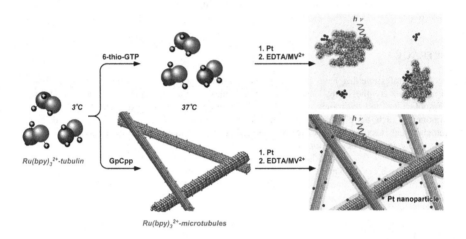

Figure 1. Control of hierarchical structure from tubulin to microtubules for photoinduced H$_2$ generation by using Ru(bpy)$_3$$^{2+}$-conjugated tubulin. EDTA = ethylenediaminetetraacetic acid, MV = methylviologen.

DISCUSSION

Ru(bpy)$_3$$^{2+}$-conjugated tubulin was prepared by amine coupling among Ru(bpy)$_3$$^{2+}$-succinimidylester and primary amine groups on microtubule surfaces.[3] To investigate the stability of the Ru(bpy)$_3$-tubulin and the Ru(bpy)$_3$$^{2+}$-conjugated microtubules in a mixture of the photochemical reaction, two types of suspensions were prepared: suspension A containing 6-thio-GTP to keep as tubulin without microtubules formation, and suspension B containing GpCpp to keep as microtubules (see Fig. 1). After increasing the suspensions temperature from 3°C to 37°C, the suspensions were mixed with EDTA, MV^{2+}, and tiopronin-Pt nanoparticles for the photoinduced H$_2$ generaton (Fig. 2a). Firstly, by detecting fluorescence from *Ru(bpy)$_3$$^{2+}$

molecules, the tubulins or the microtubules were observed. In the suspension A, tubulin flocculations over 10 μm size were mostly observed. By contrast, in the suspension B, microtubules with ~5 μm length were observed, which suggest that the Ru(bpy)$_3^{2+}$-microtubules were stable in spite of coexistance with the chemicals for photoinduced H$_2$ generaton. These results mean that the tubulin in tertiary structures easily flocculated disorderly in the high salt concentration but the microtubules in quarternary structure kept the ordered structure without flocculation. Such a difference between A and B reveals that microtubules have an ability to maintain their structure in the chemicals.

By irradiating visible light to the suspensions, the absorbance at 603 nm originating from MV$^+\cdot$ increased, and the color changed from yellow to blue (Fig. 2b). As the next forward reaction, the electrons were transmitted from MV$^+\cdot$ to Pt nanoparticles, and H$_2$ actually generated (Fig. 2c). Here, the H$_2$ generation rate of the B was 3 times higher than that of the A. This result clearly shows that the Ru(bpy)$_3^{2+}$-microtubules as the quaternary structure was more effective for the photoinduced H$_2$ generation than the flocculated tubulins. In particular, the electronic transmission from the MV$^+\cdot$ to Pt nanoparticles for the H$_2$ generation should be more effective in the microtubules suspension than the flocculated tubulins suspension. In the flocculated tubulins suspension, the Pt nanoparticles in the sparse liquid phase easily aggregated to precipitate as well as in the case of the bulk solution, or the aggregations adsorbed on the tubulin flocculations as a result of the nanoparticles' instability in the high salt concentration (see Fig. 1). In contrast, the Pt nanoparticles in microtubules network were maintained to disperse adsorbing on the microtubules network without the aggregation. In this way, the microtubules network synergistically resulted in the dispersibility of the catalytic Pt nanoparticles.

After the 12h irradiation, the suspension A became more cloudy but the suspension B kept transparent (see Fig. 2b). This result suggests that tubulin flocculated much more but the microtubules maintained the quaternary structure during the photochemical reaction. This was also confirmed by the absorption spectra baseline increased much more after the irradiation in the suspension A than in the suspension B. Thus, the flocculated tubulins were irreversibly disordered on exposure to the chemical whereas the microtubules had the resistance, and maintained the higher structure for the reaction field of the photoinduced H$_2$ generation.

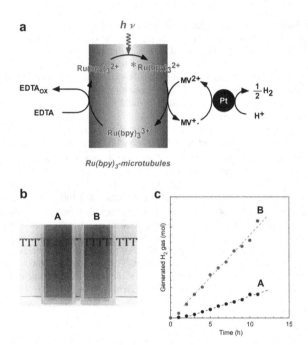

Figure 2. a) The photoinduced electronic transmission circuit on the Ru(bpy)$_3$$^{2+}$-conjugated microtubules. b) The suspensions after 12 hours irradiation. c) The photoinduced H$_2$ generation of the suspensions at 30°C. [6-thio-GTP]$_0$ = 5 mM for the A, and [GpCpp]$_0$ = 5 mM for the B; [EDTA]$_0$ = 20 mM; [MV^{2+}]$_0$ = 2.0 mM; [Pt] = 1.0 μM; BRB80 buffer (80 mM PIPES; 1 mM EGTA; 5 mM MgCl$_2$, pH6.8).

CONCLUSIONS

The Ru(bpy)$_3$$^{2+}$-conjugated microtubules were fabricated as an operating medium for photoinduced H$_2$ generation. The microtubules network synergistically caused dispersion of Pt nanoparticles in the mixture of photoirradiation. Therefore, the Ru(bpy)$_3$$^{2+}$-microtubules exhibited more efficient H$_2$ generation than the Ru(bpy)$_3$$^{2+}$-tubulins flocculation. The microtubules acted as a hierarchical medium for cooperation among the photosensitizer and the catalytic nanoparticle. In future, use of a medium with hierarchy enables many designs of multiple components arrangement, and effectively photoinduced electronic transmission toward artificial photosynthesis is envisioned.

ACKNOWLEDGMENTS

K. O. is grateful for the research fellowships of the Japan Society for the Promotion of Science for Young Scientists.

REFERENCES

1. K. Okeyoshi, R. Yoshida, Soft Matter **2009**, 5, 4118.
2. M. Castoldi, A. V. Popov, Protein Expression Purif. **2003**, 32, 83.
3. J. Peloquin, Y. Komarova, G. Borisy, Nat. Methods **2005**, 2, 299.
4. K. Okeyoshi, R. Kawamura, R. Yoshida, Y. Osada, J. Mater. Chem. B, **2014**, 2, 41.

Materials for Neural Interfaces

Mater. Res. Soc. Symp. Proc. Vol. 1621 © 2014 Materials Research Society
DOI: 10.1557/opl.2014.303

Investigating the surface changes of silicon in vitro within physiological environments for neurological application

Maysam Nezafati[1], Stephen E. Saddow[1], Christopher L. Frewin[1]
[1]University of South Florida, Department of Electrical Engineering 4202 E Fowler Ave, Tampa, FL 33620, U.S.A.

ABSTRACT

Silicon has been used as one of the primary substrates for micro-machined intra-cortical neural implants (INI). The presence of various ions in the extracellular environment combined with cellular biological activity establishes a harsh, corrosive environment in the brain for INI, and as such, a long-term implant's construction materials must be able to resist these environments. We have examined if environmental components could contribute to changes in the material, which in turn may be a contributing factor to the decreased long-term reliability in INI optimal neural recordings, which have prevented clinical use these devices for the last 4 decades. We tested silicon in artificial cerebrospinal fluid (ACSF), Dulbecco's modified eagle medium (DMEM), and H4 cells cultured within DMEM for 96 hours at 37°C as three various physiological environments to investigate the material degradation. We have observed that Si samples immersed in only DMEM and ACSF showed very minor surface alterations. However, Si samples cultured with H4 cells exhibited a large change in surface roughness from 0.24 ± 0.04 nm to 4.85 nm. The scanning electron microscope (SEM) micrographs showed the presence of pyramid shaped pits. Further characterization with atomic force microscope (AFM) verified this result and quantified the severe changes in the surface roughness of these samples. At this initial stage of the investigation, we are endeavoring to identify the cause of these changes to the Si surface, but based on our observations, we believe that the increased corrosion could be result of chemical products released into the surrounding environment by the cells.

INTRODUCTION

In our previous investigation, we observed that Si had surface damage after cell culture, but the exact source of this damage was unknown [1-3]. As Goodwin et al. in 1997 and Pocock et al. in 2001 indicated in their work that the cultured hippocampal microglial cells could release nitric oxide [4, 5]. Wink et al. in 1998 provided evidence about the role of nitric oxide in biochemistry of the neurological systems [6, 7]. Release of nitric oxide in the extracellular fluid within the brain by inflammatory cells can result in formation of nitrous and nitric ions which act as corrosive agents for silicon. ISO 10993 suggests a set of corrosion test prior to any *in vivo* application of the implantable materials and devices [8, 9]. In this experiment, we devised a method based off the ISO standard to identify the source of the surface modifications seen after cell culture. The surface morphology of (100)Si was examined using samples cultured with H4 neuroglioma cells against samples soaked in Dulbecco's modified eagle medium (DMEM) and artificial cerebrospinal fluid (ACSF).

EXPERIMENT

The samples used in this work were: a silicon wafer with the crystallographic direction of (100) (thickness of 500 μm, University wafers); a silicon wafer coated with thin layers of titanium (200 Å, 99.9%, Sigma Aldrich) and gold (1500 Å, 99.99%, Sigma Aldrich); a silicon wafer coated with thin layers of titanium (200 Å, 99.9%, Sigma Aldrich) and copper (1500 Å, 99.99%, Sigma Aldrich). The wafers were cut into 8×10 mm coupons using a diamond-coated blade on a dicing saw. The metallic coated Si coupons were used as scientific controls for cytotoxicity test. The Cu samples were the positive cytotoxicity control and Au samples were the negative cytotoxicity control. The metallic coupons were cleaned using acetone and isopropanol for 10 minutes each in ultrasound and then rinsed with deionized (DI) water, $\rho > 16$ MΩ cm. The uncoated Si coupons were cleaned with piranha and HF in addition to solvent cleaning. The piranha solution concentration was ($3:1$ H_2SO_4: H_2O_2) and samples were exposed for 5 minutes at 85°C. To remove the native oxide layer, the samples were immersed in dilute hydrofluoric acid (50:1 H_2O: HF) for 30 seconds. The cleaning stages were followed by rinsing in deionized (DI) water $\rho > 16$ MΩ cm.

The cytotoxicity test was accomplished through the direct platting of H4 Cells onto the surface of the test materials (Si), the controls (Cu, Au) and a baseline of tissue culture treated, 22 mm diameter, round polycarbonate (CT-PC) cover slides. Culture treatment allows for better cell attachment, viability and proliferation by adding oxygen species into the polymer chains. 120,000 H4 neuroglioma cells (ATCC HTB-148, American Type Culture Collection) were plated onto the coupons and incubated within a sterile environment at 37 °C, 95% relative humidity, and 5% CO_2 for 96 hours. In parallel to the cytotoxicity testing, another set of uncoated silicon samples was immersed in DMEM and ACSF and the samples were exposed to the same incubation conditions.

First, we evaluated the cellular attachment to the Si, Au and Cu samples using the fluorescent microscopy. The sample was carefully removed from the well. The dye was added to the cells to provide a fluorescent tag with which to determine the live or dead status of the cells. The fluorescent dye was consisted of 10 ml of 10 mM Phosphate buffer saline (PBS) diffused with 5 μl of 1 mg/ml calcein dye (1mg/ml in anhydrous DMSO, life technologies) to detect live cells and 7.5 μl of 2 mM ethidium homodimer-1 dye (EthD-1, 2 mM solution in 1:4 DMSO/H_2O, Invitrogen) to detect dead cells. After recording the live/dead populations of the cells, the uncoated Si samples retrieved for further surface characterization. For statistical analysis the population of live cells on the materials was normalized to the population of cells on CT-PC slide with the same surface area as the material. This normalization will help us to have the ability of comparing the results of the test materials not only with the controls but also with the other materials that were tested with the same method.

We characterized the surface of each set of Si samples at each stage of the cleaning procedure to attempt to identify the source of the surface modifications of the Si. Surface characterization consisted of qualitative evaluation through scanning electron microscopy (SEM, Hitachi S-800) at 1000X magnification, acceleration voltage of 25 kV, working distance (WD) of 5 mm and sample surface tilting angle of 0°. The gold/ palladium coating was not used for the scans. Quantitative evaluation was obtained through atomic force microscopy (AFM, Park systems Inc. XE-100) at 45×45 μm, with scanning rate of 0.5 Hz, in tapping mode using Si tips (Budget Sensors). The cleaning consisted of three stages: solvent cleaning using acetone and isopropanol (degreasing agents), piranha solution cleaning (3:1, H_2SO_4: H_2O_2) (removal of carbon containing materials), and hydrofluoric acid (1:4, HF: H_2O) (removal of silicon dioxide).

RESULTS

Figure 1 shows the fluorescent micrograph of selected samples of Cu, Au, and Si. The H4 cells distribution can be seen on the surface of the materials. A 10X objective lens was used to capture the total population of the cells on the coupons.

Figure 1, The fluorescent micrographs of the H4 Cell distribution on the surface of a) Cu (positive control), b) Au (negative control) and c) Si (test material) after culturing H4 cells in DMEM at 37°C for 96 hours. The length of scale bar is equal to 1000 μm.

Figure 2 shows the cell proliferation (%) results of Cu (positive control), Au (negative control) and the Si (test material). The average cell proliferation percentage including standard error was 0.88 ± 0.3 % for Cu, 81.71 ± 4.1% for Au, and 53.32 ± 3.1 % for Si.

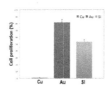

Figure 2, Cell proliferation and viability for Si, Cu and Au of H4 Cells expressed as $x \pm \sigma_M$, The live H4 cells counted were normalized to the number of live cells in the culture treated polycarbonate baseline well (CT-PC). H4 cells were cultured in DMEM at 37°C for 96 hours.

Figure 3 shows representative SEM micrographs of Si before the test, after soaking in ACSF, after soaking in DMEM, and after plating H4 cells.

Figure 3, SEM micrographs of a) untested Si, Si samples immersed at 37°C for 96 hours in b) ACSF, c) DMEM and d) DMEM + H4 Cells, in 1000X magnification with acceleration voltage of 25 kV, WD of 5 mm and tilting angle of 0°. The length of scale bars are 50 μm.

Figure 4 shows the AFM micrographs of Si before testing, after soaking in ACSF, after soaking in DMEM, and after plating H4 cells.

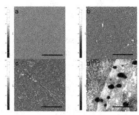

Figure 4, 45×45 μm area AFM micrographs of a) untested Si, b) Si samples immersed in ACSF, c) Si immersed in DMEM and d) Si immersed in DMEM + H4 Cells, for 96 hour at 37°C. The micrographs were normalized to a Z height interval of (+10, -10 nm) and the scale bar's length is 20 μm.

The mean root mean square surface roughness (R_q) and standard error of the Si samples for cytotoxicity samples, DMEM soaked samples and ACSF soaked samples after three stages of cleaning is presented in Table I. The surface roughness of untested Si was 0.24± 0.04 nm.

Table I, Surface roughness data extracted from AFM micrographs of Si in ACSF, DMEM and DMEM + H4 Cells after each stage of cleaning, expressed as $x \square \pm \sigma_M$.

Cleaning method	Data	Cell Media+H4 Cells	Cell Media	ACSF
Solvent cleaned	R_q (RMS)	10.31 ± 3.1 nm	4.39 ± 1.6 nm	2.66 ± 0.8 nm
Solvent + Piranha Cleaned	R_q (RMS)	6.21 ± 1.2 nm	3.37 ± 1.9 nm	2.55 ± 0.77 nm
Solvent + Piranha +HF cleaned	R_q (RMS)	4.85 ± 1 nm	0.89 ± 0.07 nm	0.76 ± 0.01 nm

DISCUSSION

Cytotoxicity tests:

The Au samples, as our negative control, showed no toxic reaction with H4 cells in cytotoxicity test, as is shown in Figure 1. The cells show appropriate attachment to the surface of the material and were quantified at 80% live cells compared to the culture treated polycarbonate cell culture baseline. The Cu sample on the other hand showed a complete toxic reaction with the H4 cells, as it can be seen in Figure 2, possessing less than 1% of the H4 cells survived. The combined negative and positive control results indicate that the cells reacted as expected during the duration of the tests. The Si samples showed an average cell proliferation of 53.32 ± 3.1 %, which in comparison to the negative control which had 30% less proliferation. Figure 1c showed an uneven distribution of cell attachment to the surface of the Si sample, which could be a result of degradation or chemical changes on the surface of the Si during the test. Silicon reacts with oxygen to form a native oxide, and in aqueous environments, this oxide can react to form silicic acid ($SiO_x (OH)_{4-2x}$). This product has been shown to be cytotoxic to cells in large local concentration in previous studies and can be in part the reason for the lower proliferation [10].

Surface degradation tests:

To separate the major factors that may have caused the surface degradation of Si, the samples were tested in three distinct environments consisting of: the presence of H4 cells in

DMEM, DMEM in absence of the cells, and ACSF, an artificial formulation of biological cerebral-spinal fluid which is suggested in ISO 10993. From figure 3a and 4a, the original surface of the silicon sample was very smooth and flat due to chemical mechanical planarization (CMP). Consequentially, no particulates or depressions can be seen on the surface of the untested material. Based on our observation from the SEM micrographs in figure 3 and AFM micrographs in figure 4, miniscule surface modifications occurred across the surface of the samples in presence of DMEM and ACSF. Before cleaning, this surface change was more significant, but was considerably reduced after cleaning the samples with piranha and HF, indicating the presence of organic particulates or solidified salts were most likely to be the source of the observed surface features. Anisotropic etching, or chemical etching, of (100) Si produces a distinctive pyramid shape due to faster etching rate of the (100) plane as compared to the (111) planar direction [11]. As the SEM and AFM micrographs did not display the presence of the typical pyramid shaped pits, we have no indication of the chemical etching of (100)Si.

Alternatively, the morphology of Si samples exposed to the presence of H4 cells possessed a significant surface modification. This surface change was also seen and increased in size through the various cleaning stages, with the largest effect noticed after HF cleaning. As HF removes SiO_2, this would indicate that the surface was partially oxidized. The qualitative results from SEM and AFM analysis verify the presence of multiple pyramid shape pits on the surface of this type of materials, as seen in figure 3d and 4d. The average pit depth was 57 ± 8 nm with average width of 4600 ± 70 nm. The deepest pit that was observed has the depth of 138 nm. Due to the absence of these features on the soaked samples, we theorize that the cells may be the most likely source of the surface etching. Inflammatory cells have been shown to release chemicals during frustrated phagocytosis to dissolve the invading material [11]. Furthermore, it has been reported that neural inflammatory cells, like microglia, produce large amounts of nitric oxide, NO, a free radical which is used in signaling and cellular defense, and hydrogen peroxide, an oxidizing agent. NO in aqueous environments can chemically react to produce numerous species, one of which is nitric acid (HNO_3), a known anisotropic etchant of Si [6, 7, 12]. These chemicals could be the source of the oxidation and subsequent etch revealed by the HF, but we will need to use alternative measurement techniques, like mass spectroscopy, to exactly quantify the chemical factors involved. Furthermore, we need to evaluate if this effect is particular to the H4 glial derived immortalized cell line, or if it is characteristic of all cells we will use immortalized mouse fibroblast, L929 and primary derived rat neurons as a comparison.

CONCLUSIONS

The (100) Si sample showed moderate cell attachment in comparison to Cu (positive control) and Au (negative control) surfaces. The lack of surface modification from the various media only exposed samples, combined with the presence of a sufficient amount of cells on the Si surface seems to indicate that chemicals extruded from the cells reacted with the Si. This conclusion is strengthened by the presence of pyramidal shaped pits, an indication anisotropic etching of (100) Si. The exact chemical characteristics released from the H4 cells which lead to this surface morphology change was not able to be determined through the present experimental processes, but we are examining the effect further through mass spectroscopy and also using other cells such as L929 mouse fibroblasts and primary rat cortical neurons.

ACKNOWLEDGMENTS

This work was sponsored by the Defense Advanced Research Projects Agency (DARPA) MTO under the auspices of Dr. Jack Judy through the Space and Naval Warfare Systems Center, Pacific Grant/Contract No. N66001-12-1-4026 - Biocompatibility of Advanced Materials for Brain Machine Interfaces. The authors would like to appreciate the assistance of NREC of USF and Rosekamp labs crew.

REFERENCES

[1]   C. L. Frewin, C. Locke, S. E. Saddow, and E. J. Weeber, "Single-crystal cubic silicon carbide: An in vivo biocompatible semiconductor for brain machine interface devices," in *Engineering in Medicine and Biology Society,EMBC, 2011 Annual International Conference of the IEEE*, 2011, pp. 2957-2960.

[2]   C. L. Frewin, M. Jaroszeski, E. Weeber, K. E. Muffly, A. Kumar, M. Peters, A. Oliveros, and S. E. Saddow, "Atomic force microscopy analysis of central nervous system cell morphology on silicon carbide and diamond substrates," *Journal of Molecular Recognition*, vol. 22, pp. 380-388, 2009.

[3]   C. Coletti, M. J. Jaroszeski, A. Pallaoro, A. M. Hoff, S. Iannotta, and S. E. Saddow, "Biocompatibility and wettability of crystalline SiC and Si surfaces," in *Engineering in Medicine and Biology Society, 2007. EMBS 2007. 29th Annual International Conference of the IEEE*, 2007, pp. 5849-5852.

[4]   J. L. Goodwin, M. E. Kehrli Jr, and E. Uemura, "Integrin Mac-1 and β-amyloid in microglial release of nitric oxide," *Brain Research*, vol. 768, pp. 279-286, 9/12/ 1997.

[5]   J. M. Pocock and A. C. Liddle, "Microglial signalling cascades in neurodegenerative disease," in *Progress in Brain Research*. vol. Volume 132, M. N.-S. B. Castellano Lopez, Ed., ed: Elsevier, 2001, pp. 555-565.

[6]   D. A. Wink, "The chemical mechanisms in regulatory, cytotoxic, and cytoprotective roles of nitric oxide.," *Abstracts of Papers of the American Chemical Society*, vol. 215, pp. U360-U360, Apr 2 1998.

[7]   D. A. Wink and J. B. Mitchell, "Chemical biology of nitric oxide: Insights into regulatory, cytotoxic, and cytoprotective mechanisms of nitric oxide," *Free Radical Biology and Medicine*, vol. 25, pp. 434-456, Sep 1998.

[8]   I. O. f. Standardization, "ISO 10993-15:2000," in *Biological evaluation of medical devices -- Part 15: Identification and quantification of degradation products from metals and alloys*, ed, 2000.

[9]   I. O. f. Standardization, "ISO 10993-14:2001," in *Biological evaluation of medical devices -- Part 14: Identification and quantification of degradation products from ceramics*, ed. Switzerland: ISO copyright office, 2001.

[10]   R. K. Iler, *The chemistry of silica. Solubility, polymerization, colloid and surface properties, and biochemistry*. New York/Chichester/Brisbane/Toronto: John Wiley & Sons, 1979.

[11]   D. M. Brown, I. A. Kinloch, U. Bangert, A. H. Windle, D. M. Walter, G. S. Walker, C. A. Scotchford, K. Donaldson, and V. Stone, "*An in vitro study of the potential of carbon nanotubes and nanofibres to induce inflammatory mediators and frustrated phagocytosis*," *Carbon*, vol. 45, pp. 1743-1756, 2007.

[12]   B. Schwartz and H. Robbins, "Chemical Etching of Silicon," *Journal of the Electrochemical Society*, , vol. 123, pp. 1903-1909, 1961.

Mater. Res. Soc. Symp. Proc. Vol. 1621 © 2014 Materials Research Society
DOI: 10.1557/opl.2014.266

Quantification of Axonal Outgrowth on a Surface with Asymmetric Topography

Elise Spedden[1] and Cristian Staii[1]
[1]Tufts University Department of Physics and Astronomy, 4 Colby St, Medford, MA 02155

ABSTRACT

Topographical features are known to influence the axonal outgrowth of neurons. Understanding what kinds of topographical features are most effective at growth cone guidance and how outgrowth responds to these structures is of great importance to the study of nerve regeneration. To this end we analyze axonal outgrowth on tilted nanorod substrates which have been shown to impart directional bias to neuron growth. We utilize the Atomic Force Microscope to characterize the surface features present on these substrates and how such features are influencing the axonal outgrowth. Additionally, using a model which considers the neuronal growth cone as an object influenced by an effective potential we determine an effective force imparted on the growth cone by the surface topography.

INTRODUCTION

The study of neuronal outgrowth on patterned surfaces is important for understanding nervous system development and repair [1, 2]. Such studies focus both on understanding the underlying sensory mechanisms behind directed neuronal outgrowth, and on optimizing biocompatible surfaces for directed growth in the field of neural tissue engineering. Directed axonal outgrowth has been studied on a variety of surfaces. Particular focus has been paid to the sensing of surface bound proteins [3], signaling molecules [4], and topographical features such as ridges [5, 6] or indentations [7]. Axons have been shown to respond to topographical patterns [5-9]. The preferred growth orientation with respect to repeating patterns is cell-type dependent [5, 6, 10], and neurons have been shown to consistently respond to variations in surface topography and micro-patterning [5-7, 9-11].

Many studies of axonal outgrowth on patterned substrates rely on symmetric patterns to impart outgrowth along one or more pattern axes [5-7, 9]. The neurons in these studies typically prefer alignments which can be characterized as parallel or perpendicular to the repeating topographical surface patterns. Similar results have been achieved on symmetric patterns of chemical rather than topographical signals [3]. Such symmetric surfaces can bias outgrowth along an axis, but cannot be used to direct neuronal growth towards a single direction, as can be achieved through electrical stimulation [12]. Asymmetric tilted nanorod surfaces, however, have been shown to impart axonal outgrowth bias towards a single dominant direction [11]. Here we utilize similar tilted nanorod substrates. We make use of the high spatial resolution of the Atomic Force Microscope (AFM) to characterize the features of these surfaces, and analyze the directed outgrowth on these surfaces of embryonic rat cortical neurons after 5 days. Additionally, we model the growth cone as an object under the influence of an effective force to determine the relative strengths of axonal outgrowth bias towards perpendicular aligned axon growth along the nanorod tilt axis.

EXPERIMENT

Film deposition and modification

The films used are nanotextured poly(chloro-p-xylylene) surfaces formed through vapor-phase polymerization and directed deposition of [2.2] paracyclophane derivatives. These films are generated as described previously in literature [11]. This results in a forest of tilted polymer nanorods. The tilt of these nanorods is then further exaggerated through directional application of force along the surface in the rod tilt direction.

Surface preparation and cell culture

The surfaces were fixed to glass slides using silicone glue and allowed to dry for 48 hours. Each surface was then rinsed with sterile water and spin-coated with 3 mL of Poly-D-lysine (PDL) (Sigma-Aldrich, St. Louis, MO) solution (0.1 mg/mL) at 1000 RPM for 10 minutes. The plates were sterilized prior to cell culture in ultraviolet light for $\geq$30 minutes.

Isolated rat cortices were obtained from embryonic day 18 rats (Tufts Medical School). The corticies were incubated in 5 mL of trypsin at 37°C for 20 minutes. The trypsin was inhibited with 10 mL of soybean trypsin inhibitor (Life Technologies). The neurons were mechanically dissociated, centrifuged, and the supernatant was removed. The cells were then resuspended in 20 mL of neurobasal medium (Life Technologies) supplemented with GlutaMAX, b27 (Life Technologies), and pen/strep. The cells were re-dispersed with a pipette, counted, and plated at a density of 6,000 cells/cm^2. Each sample was grown for five days prior to each measurement.

Fluorescence imaging and quantification of axonal outgrowth

For fluorescence imaging the live cortical samples were rinsed once with phosphate buffered saline (PBS) and then incubated for 30 minutes at 37°C with 50 nM Tubulin Tracker Green (Oregon Green 488 Taxol, bis-Acetate) (Life Technologies, Grand Island, NY) in PBS. The samples were then rinsed twice with PBS and re-immersed in fresh PBS for imaging. Fluorescence images were taken using a standard Fluorescein isothiocyanate -FITC filter: excitation/emission of 495 nm/521 nm.

Axon outgrowth was tracked using the NeuronJ plugin for ImageJ. All axons were tracked and then partitioned into 10 µm segments. The angle of each segment was measured and histograms were generated to determine the amount of outgrowth within $\pm \frac{\pi}{12}$ radians of the 0 or

π radians directions (along or against the rod tilt direction).

AFM analysis of films

All surfaces were imaged using an MFP3D AFM (Asylum Research, Santa Barbara, CA) using AC mode operation and AC 160TS cantilevers (Asylum Research, Santa Barbara, CA). Surfaces were imaged both before and after neuronal culture, and no significant change in topography was seen.

RESULTS AND DISCUSSION

The surfaces utilized in this study are nanotextured poly(chloro-p-xylylene) surfaces formed through vapor-phase polymerization and directed deposition. The resulting structure is a thin layer composed of a forest of tilted nanorods. This nanorod forest is then further tilted by the directional application of force to the surface in the rod tilt direction. Similar substrates have been characterized in literature [11]. In order to characterize the detailed topography of the tilted nanorod substrates AFM images were acquired over the substrates prior to cell seeding. Topographical images were obtained and height profile information was used to characterize the surface properties on the physical scale of an advancing growth cone.

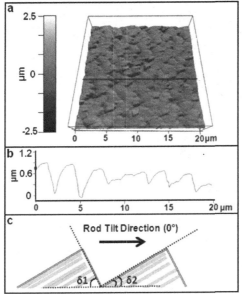

Figure 1. (a) 20x20 μm AFM topographical scan of the nanorod substrate with an asymmetric microratchet topography. The color scale indicates feature height. (b) Height profile of a line across the substrate indicated by the red line shown in (a). This height profile demonstrates the repeating ratchet topography exhibited by the substrate. (c) Schematic identifying the definitions of the angles δ1 and δ2 on the ratchet topography.

AFM analysis of the tilted nano-rod substrates reveals an emergent nanorod bundle superstructure. This morphology can be seen in the AFM scan shown in Figure 1 (a). The nanorod clump structure forms an effective surface microratchet topography, roughly imitating a repeating pattern of parallel ratchets leaning towards the nanorod tilt direction. This ratchet topography can be seen more clearly in the height profile shown in Figure 1 (b). The ratchet surfaces facing the rod direction (0°) are steeper than those facing away from the rod direction

(180°). This is characterized by the ratchet angles δ1 and δ2 shown in Figure 1 (c). For these surfaces δ1 = 40° ±10° and δ2 = 20° ± 7°.

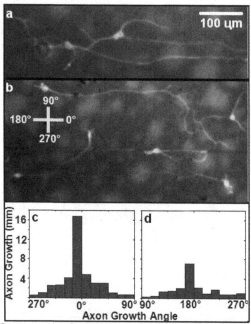

Figure 2. (a and b) Representative fluorescent images of cortical neurons grown on a nanorod substrate with asymmetric microratchet topography. The angle θ is defined with respect to the underlying topography as pointing in the ratchet direction. (c and d) Histograms of axon outgrowth with respect to θ centered on θ=0° (c) and θ=180° (d).

Cortical neurons were grown for 5 days on the substrates and axon outgrowth was quantified as a function of outgrowth angle θ. Outgrowth at θ = 0° is defined as outgrowth in the nanorod/microratchet tilt direction. Examples of axon outgrowth are shown in Figure 2 (a and b). Histograms of axon outgrowth by angle are shown in Figure 2 (c and d). We see from this figure that there are clear peaks in both directions along the nanorod tilt axis and that the peak in the nanorod/microratchet tilt direction is significantly larger than that in the opposite direction. To quantify the directional bias imparted by the topographical anisotropy we compare the total axonal outgrowth in the θ = 0° ± 90° half-plane to the outgrowth in the θ = 180° ± 90° half-plane. On our surfaces 67% of axon outgrowth length extends in the 0° ± 90° half-plane (nanorod/microratchet tilt direction), with only 33% of outgrowth occurring in the 180° ± 90° half-plane. This indicates a clear directional bias for the nanorod/microratchet tilt direction.

Within each half-plane (θ = 0° ± 90°, or θ = 180° ± 90°) we see a clear peak indicating a strong preference for perpendicular growth with respect to the parallel microratchet structure.

Analogous to our previous description for the probability distribution of axonal outgrowth velocity $p_s(v)$ [13], we consider our distributions of axonal outgrowth angles to represent the steady-state probability distribution $p(\theta)$ for an individual outgrowth segment to lie along a given angle within a half-plane. Additionally, as in our previous model [13], for each half-plane our probabilities approximate at Laplace distribution. This probability distribution yields an effective potential $V(\theta)$ via equation 1 where C represents constant terms not dependent on θ.

$$V(\theta) = -\ln[p(\theta)] * C \tag{1}$$

An effective force can thus be defined as the gradient of this effective potential ($F = \frac{\partial V}{\partial \theta}$). This effective force (with respect to θ) acts on an axon segment within a given half-plane driving it towards perpendicular alignment with the microratchet structure. This drive towards aligned growth can thus be quantified as the slope of the potential $V(\theta)$. $V(\theta)$ is plotted in Figure 3 for (a) the $\theta = 0° \pm 90°$ plane and (b) the $\theta = 180° \pm 90°$ plane.

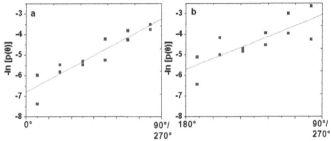

Figure 3. (a) Plot of -ln $[p(\theta)]$ vs. θ for $\theta = 0° \pm 90°$. (b) Plot of -ln $[p(\theta)]$ vs. θ for $\theta = 180° \pm 90°$. The values for $p(\theta)$ are obtained from the axon outgrowth distributions for cortical neurons grown for 5 days on a surface with asymmetric topography. -ln $[p(\theta)]$ represents the effective potential generated by this topography within each 180° half-plane on the surface.

The slope of the potential within the $0° \pm 90°$ half-plane ($F_{180} = 6.4 \pm 0.6$) is slightly steeper than that in the $\theta = 180° \pm 90°$ half-plane ($F_0 = 4.8 \pm 1.1$) indicating a slightly larger effective force driving aligned growth in the nanorod/microratchet tilt direction.

CONCLUSIONS

Neurons grown on the nanorod substrate show directed growth. A bias is seen for growth in the direction of the microratchet/rod tilt. Furthermore, a strong preference is shown for growth perpendicular to the microratchet structures. Analysis of this outgrowth as being driven towards perpendicular alignment via an effective force indicates that this force is slightly stronger in the nanorod/microratchet tilt direction.

Integrins are a part of the focal complexes which mediate adhesion and the transduction of mechanical forces within moving cells or cell structures such as growth cones. Integrin-based focal complexes mature into stable focal contacts upon the application of force [14] with the strength of the resulting contact being force-dependent. These focal contacts are important for adhesion and signal transduction during cell migration or outgrowth. Surface topography alters the strength and duration of focal contact [15] with substrates composed of steep pillars resulting in aligned growth which was faster than growth on a flat substrate. This difference was due to the fact that the duration of focal contacts were longer on pillars than on the flat surface.

The tendency for axonal outgrowth to align perpendicular to a series of ridges has been established previously in literature [5, 6]. Additionally, steep topographical features are shown to increase and direct cell migration based on changes in focal contact maturation [15]. Our results suggest that the angle of surface features with respect to the substrate plays a role in directed outgrowth of neurons on patterned surfaces. We hypothesize that this result may be due to substrate contact angle dependent differences in the formation and maturation of focal contacts. In our case, the steeper angled side of the micro-ratchet generates both more outgrowth and stronger alignment than the shallow angled side. This results in two peaks of axonal outgrowth, a major peak in the nanorod/microratchet direction, and a minor peak opposing this direction.

ACKNOWLEDGEMENTS

The authors thank Dr. Melik Demirel and Dr. Koray Sekeroglu for providing the substrates used in this study. We thank Dr. Steve Moss's laboratory at Tufts Center of Neuroscience for providing embryonic rat brain tissues under Tufts University approval animal care protocols.

REFERENCES

1.   Z. Wen and J. Q. Zheng, Current opinion in neurobiology **16** (1), 52-58 (2006).
2.   A. B. Huber, A. L. Kolodkin, D. D. Ginty and J. F. Cloutier, Annual review of neuroscience **26**, 509-563 (2003).
3.   P. Clark, S. Britland and P. Connolly, Journal of cell science **105 (Pt 1)**, 203-212 (1993).
4.   T. E. Kennedy, T. Serafini, J. R. de la Torre and M. Tessier-Lavigne, Cell **78** (3), 425-435 (1994).
5.   A. Rajnicek, S. Britland and C. McCaig, Journal of cell science **110 (Pt 23)**, 2905-2913 (1997).
6.   A. Rajnicek and C. McCaig, Journal of cell science **110 (Pt 23)**, 2915-2924 (1997).
7.   D. Y. Fozdar, J. Y. Lee, C. E. Schmidt and S. Chen, Int J Nanomedicine **6**, 45-57 (2011).
8.   Y. W. Fan, F. Z. Cui, S. P. Hou, Q. Y. Xu, L. N. Chen and I. S. Lee, Journal of neuroscience methods **120** (1), 17-23 (2002).
9.   F. Johansson, P. Carlberg, N. Danielsen, L. Montelius and M. Kanje, Biomaterials **27** (8), 1251-1258 (2006).
10.   R. M. Smeal and P. A. Tresco, Experimental neurology **213** (2), 281-292 (2008).
11.   R. Beighley, E. Spedden, K. Sekeroglu, T. Atherton, M. C. Demirel and C. Staii, Applied physics letters **101** (14), 143701 (2012).
12.   N. Patel and M. M. Poo, The Journal of neuroscience : the official journal of the Society for Neuroscience **2** (4), 483-496 (1982).
13.   D. J. Rizzo, J. D. White, E. Spedden, M. R. Wiens, D. L. Kaplan, T. J. Atherton and C. Staii, Physical Review E **88** (4), 042707 (2013).
14.   D. Riveline, E. Zamir, N. Q. Balaban, U. S. Schwarz, T. Ishizaki, S. Narumiya, Z. Kam, B. Geiger and A. D. Bershadsky, The Journal of cell biology **153** (6), 1175-1186 (2001).
15.   M. T. Frey, I. Y. Tsai, T. P. Russell, S. K. Hanks and Y. L. Wang, Biophysical journal **90** (10), 3774-3782 (2006).

Mater. Res. Soc. Symp. Proc. Vol. 1621 © 2014 Materials Research Society
DOI: 10.1557/opl.2014.267

Improved Biphasic Pulsing Power Efficiency with Pt-Ir Coated Microelectrodes

Artin Petrossians[1, 2], Navya Davuluri[3], John J. Whalen III[2], Florian Mansfeld[1], James D. Weiland[2, 3]

[1] Mork Family Department of Chemical Engineering and Materials Science,
[2] Department of Ophthalmology,
[3] Department of Biomedical Engineering,
University of Southern California, Los Angeles, California, USA

ABSTRACT

Neuromodulation devices such as deep brain stimulators (DBS), spinal cord stimulators (SCS) and cochlear implants (CIs) use electrodes in contact with tissue to deliver electrical pulses to targeted cells. In general, the neuromodulation industry has been evolving towards smaller, less invasive devices. Improving power efficiency of these devices can reduce battery storage requirements. Neuromodulation devices can realize significant power savings if the impedance to charge transfer at the electrode-tissue interface can be reduced. High electrochemical impedance at the surface of stimulation microelectrodes results in larger polarization voltages. Decreasing this polarization voltage response can reduce power required to deliver the current pulse. One approach to doing this is to reduce the electrochemical impedance at the electrode surface. Previously we have reported on a novel electrochemically deposited 60:40% platinum-iridium (Pt-Ir) electrode material that lowered the electrode impedance by two orders of magnitude or more.

This study compares power consumption of an electrochemically deposited Pt-Ir stimulating microelectrode to that of standard Pt-Ir probe microelectrode produced using conventional techniques. Both electrodes were tested using *in-vitro* in phosphate buffered saline (PBS) solution and *in-vivo* (live rat) models.

INTRODUCTION

Neuromodulating implants are used to treat a variety of neurological disorders, including deafness, movement disorders (Parkinson's), and chronic back pain [1-4]. Recently, implantable stimulators have been approved for use for blindness in the US [5]. Obsessive Compulsive Disorder [6], treatment-resistant depression [7], Alzheimer's disease [8], and migraine [9, 10] are emerging targets for implants. Electrodes are the critical interface between implants and the surrounding neural tissue. An implantable neurostimulator's efficacy, efficiency, longevity, precision, and the fabrication cost are impacted by the choice of electrode technology used. The evolution of these devices continues to move towards smaller sized and greater quantity of electrodes per device. This evolution is driven by the need for more precise and complex stimulus patterns, to avoid side effects seen in brain stimulation and to enable complex sensory input as in visual prostheses.

Electrode size reduction increases the electrochemical impedance of the electrode-tissue interface, reducing power efficiency. For chronic stimulation/pacing devices, where the same amount of charge must be delivered across a smaller interface, the impedance increase results in increased power consumption [11]. It is therefore desirable to minimize electrochemical impedance to maximize efficiency of signal transmission.

Recently, we reported on the development of a novel approach to prepare low impedance electrodes by electrochemical co-deposition of platinum and iridium. This approach exploits the simplicity and efficiency of the electrochemical reduction process to form conductive coatings of platinum and iridium with predetermined compositions and morphologies [12]. Co-deposition of the two metals results in a coating with improved charge transfer properties at the electrochemical interface, and improved mechanical behavior compared to the mechanical properties of the component metals. The concept of improving charge transfer properties of stimulation electrodes through application of low impedance coatings is not new [13-17], and the primary properties evaluated include composition, charge capacity, and mechanical properties, among others.

In this study, a low-impedance platinum-iridium alloy coating was electrochemically deposited onto a platinum-iridium probe-style disk electrode. This electrode was then surgically placed on the retina of a rat via a small sclerotomy (i.e. incision through the eye wall). We investigated the effect of the coating on stimulus pulse power consumption when used to electrically stimulate visual percepts in the rat retina. Performance was compared against a similar electrode but without the platinum-iridium coating applied. The goal was to determine if any significant power savings was gained using the platinum-iridium coating on a stimulation electrode.

EXPERIMENTS

Stimulation Electrodes
Concentric bipolar Pt-Ir electrodes (Model CBDFG74, FHC, Bowdoin, ME) with a flat, circular tip were used as stimulation electrodes. The diameter of the inner pole was 75 μm. The surrounding ring electrode (the outer pole) was not used in this study. Rather, the electrode was used in a monopolar configuration with the inner pole used for stimulation. A large surface area platinum needle inserted in the skin adjacent to the nose was used as the return electrode. The stimulating electrode was mounted inside a 1 mL syringe with the probe tip extending through the luer lock aperture, to protect the probe from bending and damage during handling.

Modification and Characterization of Stimulation Electrodes
Stimulation electrodes were modified by electrochemically depositing a high surface area platinum-iridium coating. The method of deposition has been reported elsewhere [12]. Coatings were applied using the same deposition conditions on all electrodes to ensure that similar morphologies and compositions were employed in all studies.

Structural and morphological characterization of the uncoated and Pt-Ir coated electrode surfaces were performed using a ZEISS 1550VP field emission scanning electron microscope (Carl Zeiss Inc., Oberkochen, Germany). Chemical compositions of the coatings were characterized using energy dispersive spectroscopy (EDS) (Oxford EDS - HKL EBSD). EDS analyses of the MEAs were performed at three separate locations for each electrode. EDS was used to investigate the existence and percentage of the Pt Ir elements only.

Electrochemical characterization was performed on coated and uncoated electrodes at baseline using electrochemical impedance spectroscopy (EIS). EIS measurements were performed using a Gamry Reference 600™ potentiostat (Gamry Inc. Warminster, PA) in a three-electrode setup.

Animals

Long Evans rats (postnatal day P90 – P120, n = 8) were used in the study. Animals were randomly assigned to two (2) experimental groups. Group 1 used a modified stimulation electrode. Group 2 used an unmodified stimulation electrode. Animals were housed in covered cages and fed a standard rodent diet *ad libitum* while kept in a 12:12-hour light: dark cycle animal facility. All experimental procedures were approved by the Institutional Animal Care and Use Committee (IACUC) at the University of Southern California.

Surgical Procedures

All rats underwent an ocular surgery procedure to place a stimulating electrode on the retinal surface of one eye. The stimulation electrode was attached to a single-axis linear translational micromanipulator (Model NT33-475, Edmund Optics, Barrington, NJ) on a magnetic based articulating arm to allow for careful manipulation into and out of position.

All surgeries were performed under general anesthesia induced by intraperitoneal/intramuscular injection of a cocktail of ketamine (100 mg/kg; Ketaset, Fort Dodge Animal Health, Fort Dodge, IA) and xylazine (100mg/kg; X-Ject SA, Butler, Dublin, OH) and maintained by mask delivery of sevoflurane (1% in 100% O2), flowed throughout the entire experiment. Pulse and oxygen saturation were monitored during the surgical procedures. Body temperature was maintained at 37°C with a self-regulated heating blanket (model 50-7053-F; Harvard Apparatus, Holliston, MA). Animals were euthanized after each experiment.

Stimulation Electrode Insertion

The surgical procedure to insert a stimulation electrode into the rat eye was reported in previous work [18, 19]. The left eye was dilated with a few drops each of 1% tropicamide (Tropicacyl, Akorn, Buffalo Grove, IL) and 2.5% phenylephrine (AK-Dilate, Akorn). The dilated eye was proptosed using a small piece of a surgical glove with an aperture created for the eye. The cornea was slightly flattened by placing a glass coverslip covered with gel (Goniosol, Gonak) overtop, thus allowing focused viewing of the fundus through an operating microscope. A scleral incision was made using a 25-guage needle near the limbus. The needle was inserted at a 45° angle with respect to the scleral surface in order to avoid damaging the lens. The stimulation electrode was inserted through the incision site along the path made by the needle. The electrode was positioned in the ventral temporal quadrant without contacting the retina.

Electrical Stimulation

Charge-balanced, cathodic first, biphasic current pulses were delivered to the epiretinal surface using an A-M systems converter (model 2200) that was driven by a voltage pulse from a programmable analog output card (DataWave Technologies, Berthoud, CO). Pulses were delivered in trains of four, with each pulse in the train having the same pulse duration, but with each subsequent pulse having a 10 nC larger total charge delivered. The amplitudes of the pulses were chosen such that the charge in the cathodic phase for the four pulses were: 20 nC, 30 nC, 40 nC and 50 nC in succession. Two different pulse durations were used (0.5 ms and 1.0 ms). Thus 8 test conditions were applied to the retina per experiment. An interphase interval (t = 100 μs) was used with all biphasic pulses. The stimulus pulses were delivered at 0.2 Hz for all pulse durations. Table 1 organizes the 8 different pulse conditions by duration and pulse amplitude. Twenty-five (25) total pulses of each type were delivered.

Table 1. List of Different Pulse Test Conditions.

0.5 ms Tests	Current (µA)	Charge (nC)	1.0 ms Tests	Current (µA)	Charge (nC)
1	40	20	5	20	20
2	60	30	6	30	30
3	80	40	7	40	40
4	100	50	8	50	50

Power Consumption Analysis

Power consumption associated with driving biphasic current pulses through stimulation electrodes was computed using equation 1.

$$P = \frac{1}{T}(V_{max} * \int I(t)dt) \quad (1)$$

Where $I(t)$ is the applied current measured at the electrode interface; T is the pulse duration, and dt is the step size in time. V_{max} is the maximum voltage measured from the resulting polarization voltage curve. V_{max} is a scalar value, because devices that use a current source to stimulate require a fixed voltage at least equal to the maximum voltage anticipated during use. Additional background and details regarding this approach to power estimation have been reported [11].

Power consumption associated with signal transmission through coated electrodes was compared to power consumption through uncoated electrodes, to determine if any significant savings were gained by application of the coating. A student's t-test was used to determine the statistical significance of power consumption data. Power values from the coating were deemed significantly lower than those from the standard Pt-Ir electrode, if the p-value was less than 0.05. Power values were also analyzed as a function of biphasic pulse charge level, to assess the dependence of power savings on charge level delivered.

RESULTS & DISCUSSION

Electrode Preparation, Characterization & Placement

High surface-area Pt-Ir coating was electrodeposited on the Pt-Ir microelectrodes, illustrated in Figure 1. SEM micrographs of the surface morphologies of Pt-Ir coated microelectrode, e.g. Figure1 (inset, right), provided visual confirmation of the increase in Pt-Ir film real surface area while maintaining the original geometric area. Quantitative compositional analysis performed on the electrodeposited Pt-Ir films by EDS showed an average chemical composition of about 60:40% Pt:Ir (molar fraction) in the electroplated samples, Figure 1 (right). Other studies have shown that 60-40% Pt-Ir alloys have the largest electroactive surface area and consequently, the lowest charge transfer resistance [20].

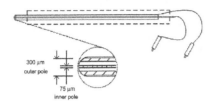
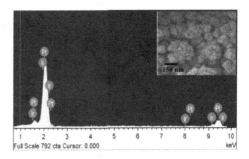

Figure 1. (Left) Diagram of the bipolar platinum-iridium stimulating electrode illustrating the probe tip (inserted into the eye) at left and the two lead connectors at right. (Right) EDS spectral analysis showing the mixed Pt and Ir composition of electrodeposited coating. The inset (top right) is an SEM micrograph of the high surface-area Pt-Ir coating.

EIS data for the uncoated and Pt-Ir coated microelectrode in phosphate buffered saline (PBS) solution are shown in Bode-plot format in Figure 2 (phase angles not shown) in which the logarithm of the impedance modulus |Z| are plotted as a function of the logarithm of the applied signal frequency, f. At high frequency (greater than approximately 1 kHz), impedance magnitudes for the coated electrode show resistive behavior representing the solution resistance. At lower frequencies, the impedance for the electroplated microelectrode was reduced by one to two orders of magnitude compared to that of the uncoated electrode.

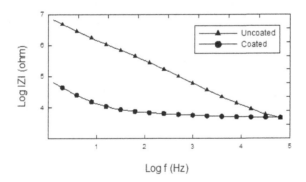

Figure 2. Comparison of impedance spectra between uncoated and Pt-Ir electrodes in PBS.

Power Consumption Comparison

Figure 3 (left) is a photograph of the *in vivo* test setup, showing the rat cornea in the central part of the figure, with stimulation electrode inserted from the right side, into the posterior camber with its tip against the posterior wall of the eye (against the retina). As mentioned, a surgical glove was used to proptose the eye, and is seen around the perimeter here. A representative biphasic current pulse driven at the stimulation electrode is inset in the corner of Figure 3 (right inset).

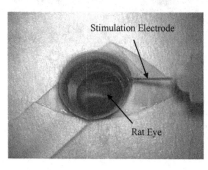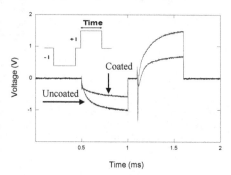

Figure 3. (Left) Optical photograph of stimulation electrode inside the rat eye, (inset, top left) biphasic current waveform, and (right) comparison of the voltage transients in response to the applied current on the uncoated and Pt-Ir coated electrodes.

Figure 3 (right) shows two representative polarization curves recorded at the stimulation electrode surface: one from a coated electrode and the other from an uncoated electrode. Biphasic current pulses applied to the uncoated electrode yielded a characteristic polarization curve, with a masked iR solution resistance step as has been reported elsewhere [13, 17]. The solution resistance segment was followed by a longer polarization period (*ca.* $t = 0.6$ to 1.0 ms). At 1.0 ms the applied current is cut off for 100 ms, then an equal but opposite anodic current pulse is applied, resulting in a voltage transient in the positive direction.

The same current pulse applied to the Pt-Ir-coated electrode resulted in significantly less voltage polarization. Generally, the voltage transient of the coated electrode showed a smaller increase in polarization over the pulse interval, and the shape of the polarization transient was more linear and less parabolic. These results were consistent with other reported comparisons of electrodes with different capacitances [17].

Figure 4 shows power consumption data plotted as a function of pulse current amplitude. Data from both 0.5 ms and 1 ms duration pulses are shown. Consistently, for all eight test pulses used, the coated electrodes showed a significantly lower average power consumption, as compared to the uncoated electrodes (p-value < 0.05 for all test cases). Shorter pulse durations ($t = 0.5$ ms) consumed more power to drive the stimulus pulse than longer pulse durations of equivalent charge level.

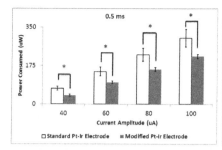 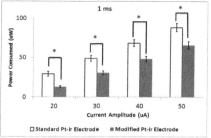

Figure 4. Power consumed to generate biphasic stimulation pulses is shown as a function of stimulus current amplitude. Data for 0.5 ms duration pulses is shown on the left and for 1.0 ms pulses is shown on the right. Error bars indicate standard error. * indicates statistical significance (p-value < 0.05).

Figure 5 compares power savings between 0.5 ms and 1.0 ms duration pulses as a function of charge delivered per half pulse. Average power consumption values for each of the eight different current pulses used are reported in Table 2. In all scenarios tested, the Pt-Ir coated electrode yielded an average 25% improvement in power consumption or greater, versus the uncoated electrode.

Comparing across test conditions, greater power savings was observed with lower total charge delivery. In fact, as large as a 55% average power savings was observed with 20 nC charge delivery at 1ms. This amount of charge is 5X threshold, as measured in experiments using this same style electrode [21]. Clinical thresholds are generally higher, but well placed electrodes can achieve thresholds in this range [22]. In chronic stimulation applications, this could lead to significant power consumption savings. This is consistent with theory. Charge transfer during electrical stimulation employs different mechanisms to reversible transfer charge from the electrode to the tissue, and vice versa. Double layer charging, having the shortest time constant, is the fastest reversible charging mechanism. Ion surface adsorption onto the electrodes and electrode metal oxidation state transitions both also contribute to reversible charge transfer, but have longer time constants. As a result, these two processes delay charge transfer enabling accumulation of some charge at the interface, leading to polarization. It is likely these longer time constant processes are lowering the power efficiency at the larger charge levels.

It is interesting to note that in the small sample size tested here, the dependence of power efficiency on pulse duration disappeared with increasing charge level. At 40 nC and greater, the power efficiency equaled 25-30% independent of pulse duration. More work is required to see if this trend is maintained at greater charge levels or if power savings continue to decrease.

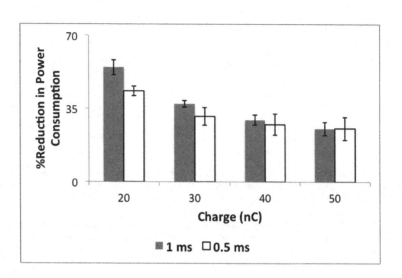

Figure 5. A comparison of % reduction in power consumption between 0.5 ms duration pulses and 1.0 ms duration pulses shown as a function of charge with the cathodic phase of the biphasic stimulation pulse. Error bars indicate standard error.

Table 2. Average power consumption data is listed for all the 8 stimulation modes.

Test	Average Power Consumed per Phase (µW)		Coating Power Savings (%)
	Uncoated	PtIr Coated	
1	71.9 ± 9.4	40.4 ± 4.5	43.8 ± 2.33
2	148.3 ± 19.3	99.4 ± 6.1	32.9 ± 4.22
3	224.6 ± 29.7	158.4 ± 7.8	29.5 ± 5.10
4	301.0 ± 40.1	217.4 ± 9.5	25.5 ± 5.47
5	29.3 ± 3.1	13.1 ± 1.2	55.4 ± 3.62
6	48.9 ± 3.5	30.7 ± 2.2	37.4 ± 1.57
7	68.7 ± 4.3	48.3 ± 3.3	29.7 ± 2.41
8	88.4 ± 5.1	65.9 ± 4.5	27.6 ± 3.20

CONCLUSIONS

1. PtIr coatings were successfully applied to probe-style microelectrodes, more specifically to circular disk PtIr stimulation. Electrochemical impedance spectroscopy confirms a marked improvement in electrochemical performance in vitro.

2. In an in vivo (rat) electrical stimulation model of the retina, PtIr coated electrodes improved power efficiency of signal transmission by 25% to 55% over use of state-of-the-art conventional electrode materials, depending on the specific pulse parameters used. Significant power savings were achieved with the coated electrodes vs. the uncoated electrodes, over the entire range of charge values (i.e. current pulse amplitudes and durations) studied.
3. Power efficiency decreases with increasing charge level delivered.
4. At lower charge levels, pulse duration affects efficiency. Specifically, longer duration pulses were more efficient at delivering charge compared to shorter pulses. This effect disappears as charge level increases.

REFERENCES

1. G.E. Loeb, *Ann. Rev. Neurosci.*, **13** (1990).
2. C.T. Tan, B. Guo, B. Martin and M. Svirsky, *J. Acoustical Soc. Amer.*, **131** (2012).
3. M. C. Rodriguez-Oroz, J. A. Obeso, A. E. Lang, J.-L. Houeto, P. Pollak, S. Rehncrona, J. Kulisevsky, A. Albanese, J. Volkmann, M. I. Hariz, N. P. Quinn, J. D. Speelman, J. Guridi, I. Zamarbide, A. Gironell, J. Molet, B. Pascual-Sedano, B. Pidoux, A. M. Bonnet, Y. Agid, J. Xie, A.-L. Benabid, A. M. Lozano, J. Saint-Cyr, L. Romito, M. F. Contarino, M. Scerrat, V. Fraix and N. Van Blercom, *Brain*, **120**, 10 (2005).
4. T. Cameron, *J. Neurosurg. Spine*, **100**, 3 (2004).
5. *http://www.sciencedaily.com/releases/2013/02/130214111112.htm.*
6. S.L. Rauch, D.D. Dougherty, D. Malone, A. Rezai, G. Friehs, A.J. Fischman, N. M. Alpert, S. N. Haber, P H. Stypulkowski, M. T. Rise, S.A. Rasmussen, and B.D. Greenberg., *J. Neurosurgery*. **104**, 4 (2006).
7. H.S. Mayberg, A.M. Lozano, V. Voon, H.E. McNeely, D. Seminowicz, C. Hamani, J.M. Schwalb and S.H. Kennedy, *Neuron*. **45**, 5 (2005).
8. A.W. Laxton, D.F. Tang-Wai, M.P. McAndrews, D. Zumsteg, R. Wennberg, R. Keren, J. Wherrett, G. Naglie, C. Hamani, G.S. Smith and A.M. Lozano, *Ann Neurol*. **68**, 4 (2010).
9. M. Leone, A. Franzini, G. Broggi and G. Bussone, *Neurol. Sci.* **24**, 2 (2003).
10. M. Leone, *Lancet Neurology*, **5**, 10 (2006).
11. S.K. Kelly, J.L.Wyatt Jr., *IEEE Trans. Biomed. Circuits and Systems*, **5**, 1 (2011).
12. A. Petrossians, J.J. Whalen, J.D. Weiland and F. Mansfeld *J. Electrochem.Soc.* **5**, 158 (2011).
13. S.F. Cogan, *Ann. Rev. Biomed. Eng.*, **10** (2008).
14. J.D. Weiland, D.J. Anderson and M.S. Humayun, *IEEE Trans. Biomed. Eng.*, **49**, 12 (2002).
15. J.J. Whalen, J. Young, J.D. Weiland and P.C.Searson, *J. Electrochem Soc.*, **135** (2006).
16. D. Zhou and R. Greenberg. "Microelectronic Visual Prostheses," *Implantable Neural Prostheses 1. Devices and Applications, Volume 1. Biological and Medical Physics, Biomedical Engineering*, ed. D. Zhuo and E. Greenbaum (Springer, 2009) pp. 1-42.
17. S. Venkatraman, J. Hendricks, A.A. King, A.J. Sereno, S. Richardson-Burns, D. Martin and J.M. Carmena, *IEEE Trans. Neural Sys. Rehab. Eng.*, **19**, 3 (2011).
18. L. Colodetti, J.D. Weiland, S. Colodetti, A. Ray, M.J. Seiler, D.R. Hinton and M.S. Humayun, *Exp. Eye Res.* **85** (2007).
19. Ray, L. Colodetti, J.D. Weiland, D.R. Hinton, E. Lee and M.S. Humayun, *Brain Res.* **1255** (2009)
20. P. Holt-Hindle, Q. Yi, G. Wu, K. Koczkur and A. Chen, *J. Electrochem. Soc.*, **155** (2008).
21. L.H. Chan, E. Lee, M.S. Humayun and J.D. Weiland, *J. Neurophys.* **105**, 6 (2011).
22. C. de Balthasar, S. Patel, A. Roy, R. Freda, S. Greenwald, A. Horsager, M. Mahadevappa, D. Yanai, M. J. McMahon, M.S. Humayun, R.J. Greenberg, J.D. Weiland and I. Fine, *Invest. Ophthalmol. Vis. Sci.* **49**, 6 (2008).

Mater. Res. Soc. Symp. Proc. Vol. 1621 © 2014 Materials Research Society
DOI: 10.1557/opl.2014.275

Atomic Layer Deposited Al₂O₃ and Parylene C Bi-layer Encapsulation for Utah Electrode Array Based Neural Interfaces

Xianzong Xie[1], Loren W. Rieth[1], Rohit Sharma[1], Sandeep Negi[1], Rajmohan Bhandari[2], Ryan Caldwell[3], Prashant Tathireddy[1], and Florian Solzbacher[1,3]
[1]Electrical and Computer Engineering, University of Utah, Salt Lake City, UT, 84112 U.S.A
[2]Blackrock microsystems, Salt Lake City, UT, 84108 U.S.A
[3]Department of Bioengineering, University of Utah, Salt Lake City, UT, 84112 U.S.A

ABSTRACT

Long-term functionality and stability of neural interfaces with complex geometries is one of the major challenges for chronic clinic applications due to lack of effective encapsulation. We present an encapsulation method that combines atomic layer deposited Al₂O₃ and Parylene C for encapsulation of biomedical implantable devices, focusing on its application on Utah electrode array based neural interfaces. The alumina and Parylene C bi-layer encapsulated wired Utah electrode array showed relatively stable impedance during the 960 equivalent soaking days at 37 °C in phosphate buffered solution. For the bi-layer coated wireless neural interfaces, the power-up frequency was constantly ~ 910 MHz and the RF signal strength was stably around -73 dBm during equivalent soaking time of 1044 days at 37 °C (still under soak testing).

INTRODUCTION

Implantable systems for long-term clinic trials require chronic implantations able to perform their intended functionalities for years or decades, in order to reduce surgical risks and generate levels of efficacy that justifies the risks associated with the implantation. Hermetic and thin-film based encapsulation are the commonly used methods to protect the device from physiological environment. Device miniaturization and electromagnetic power raise new challenges for traditional hermetic encapsulation. Thus, thin film based encapsulation have been widely employed. Compared with metal cans and lids based hermetic encapsulation, thin film based encapsulation takes less space, can handle feedthroughs easily, and is more economic and easier for mass production.

Implantable neural interfaces have been widely developed and also used to diagnose and treat neural disorders in both research and clinical applications. The Utah electrode array (UEA) is a well-developed and FDA-cleared example of this technology for stimulating and recording multiple neurons simultaneously with good selectivity[1]. Parylene C has been widely used as coating material for biomedical devices [2, 3] due to attractive properties including chemically inertness, low dielectric constant (ε_r=3.15), high resistivity ($\Box 10^{15}$ Ω·cm) and relative low water vapor transmission rate (WVTR) 0.2 g·mm/m²·day. Failure of Parylene encapsulation has also been reported [4] due to moisture diffusion and interface contamination. To overcome the condensation of moisture around interface contaminants, a highly effective moisture barrier can be introduced between the neural interface and Parylene film. Atomic layer deposited (ALD) Al₂O₃ has WVTR at the order of ~ 10^{-10} g·mm/m²·day [5]. The alumina-Parylene C bilayer encapsulation has demonstrated excellent insulation performance on planar interdigitated electrode (IDE) test structures for years of equivalent lifetime in accelerated soak testing[6-9]. However, the complex geometry, different materials and surfaces, and additional processing steps (oxygen plasma etching, BOE etching, wire bonding) involved in neural interfaces are not fully represented in IDE test structures and therefore might severely affect the actual lifetime of the bilayer encapsulated neural interfaces.

EXPERIMENT DETAILS

Wired and wireless UEA-based neural interfaces were used to evaluate the alumina and Parylene C bilayer encapsulation performance from two different aspects: long-term impedance stability, and long-term wireless signal strength and frequency stability. The fabrication details of UEAs are described elsewhere[10]. Wired UEAs were used to evaluate the electrode impedance stability over time. UEAs were wire bonded (West Bond, Inc.) to a 96-channel TDT connector using 1 mil insulated gold wire with a wirebundle length of 10 cm for long-term tip impedance measurements. The performance of the encapsulation was further tested by using wireless integrated neural interfaces, and soaking these in PBS under accelerated conditions. The ability to power the devices inductively, and the associated telemetry frequency on power-up, and the RF signal strength were used as sensitive metrics for the encapsulation performance and fluid ingress. The details of the chip design, fabrication, characterization, and system integration were reported elsewhere[11]. The fully integrated wireless INI is shown in figure 1.

Figure 1. Utah array based fully integrated wireless neural interfaces, with flip-chip bond INI-R6 and gold coil for inductive powering.

52 nm of Al_2O_3 was deposited by plasma-assisted (PA) ALD on integrated neural interfaces at a substrate temperature of 120 °C, which is within the thermal budget for the materials for the two array variants used. Details of the deposition process have been previously reported [6, 7, 9]. A 6-μm thick Parylene-C layer was deposited by chemical vapor deposition on top of Al_2O_3 as the external coating layer. For wired neural interfaces, the connectors were covered with aluminum foil to avoid coating the contact pads on the connectors. The encapsulation must be removed from the active tip electrodes sites for neural recording and stimulation. A hybrid method using a combination of laser ablation and O_2 RIE was utilized to etch Parylene C layer and buffered oxide etch was used to remove the thin alumina film [12, 13].

Wired arrays were used for long-term impedance measurements, and were soaked in 1× PBS at 57 °C for accelerated lifetime testing. The estimated aging factor (Q) was 4, based on a broadly recognized trend in accelerated aging of a doubling reaction kinetics for each 10 °C increase in reaction temperature[14]. For wireless neural interface testing, the arrays were fully submerged in 6-ml glass vials filled 1× PBS solution at 57 ± 0.5 °C in water baths. The wireless neural interfaces were powered by a customized inductive power board at 2.765 MHz that has been previously reported[11]. The presence of the 900-MHz ISM-band telemetry signal, the frequency of that signal on startup, and RF signal strength from INIR-6 chip were monitored using the custom receiver board interfaced through Matlab and with a spectrum analyzer [15].

RESULTS AND DISCUSSION

Wired Utah electrode arrays

Impedance for three wired arrays (N=3) were measured at 1 kHz using 10-mV sine wave. As shown in figure 2, tip impedances were found to range from 30 to 100 kΩ for most electrodes, with a median impedance of 60 kΩ, which are good for neural interface applications, and consistent with previously reported data [16]. The impedance of alumina and Parylene coated UEAs stayed almost the same during equivalent soaking time of first 120 days at 37 °C (non-accelerated conditions), indicating good insulation of individual electrodes. Impedance for Parylene-only control samples continuously dropped within a few weeks to 3 months[17]. Table 1 compares the median of tip impedance for Parylene-only and bilayer coated UEAs. For the Parylene-only condition, the median tip impedance dropped from 81.9 kΩ to 40.5 kΩ within 3 days of soak testing. The significant impedance drop is most likely due to water ingress and degradation of the Parylene coating. For alumina and Parylene bilayer coating, the median of tip impedance increased slightly from 61.1 kΩ to 73.8 kΩ within 3 days. After 960 days of soaking, SEM image (figure 3) confirmed the etching of exposed silicon at the electrode tips, and undercutting of the tip metallization, which contributed to increased impedance. This process occurred in the Parylene-only coating as well, and the impedances were found to decrease, which clearly suggests dramatically better insulation performance of the bilayer encapsulation.

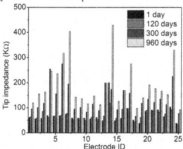

Figure 2. Electrode impedance of alumina and Parylene bilayer coated wired arrays over time. Median impedance was 60 kΩ and increased ~ 2.5 times after 960 days of soak testing in PBS.

Table 1. The median impedance for Parylene coated UEA and alumina and Parylene bilayer coated UEA for 3 days of soak testing in PBS.

Soak time	Median impedance for Parylene coated UEA (kΩ)	Median impedance for bilayer coated UEA (kΩ)
1 day	81.9	61.1
3 days	40.5	73.8

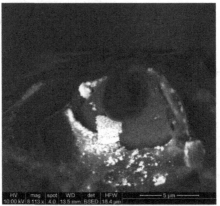

Figure 3. SEM micrograph of electrode tip after 960 days of soak testing at 37 °C.

Wireless Utah electrode arrays

Three wireless integrated neural interface (INI) (N=3) devices were soaked at 57 °C in PBS for 261 days, equivalent soak time of 1044 days at 37 °C and are still under soak testing to investigate the long-term reliability of alumina and Parylene C coated wireless INI devices. The INI device was about 8 mm away from the power coil, and the device was powered up only during testing. The presence of the signal, the startup frequency, and the RF signal strengths of the INI device at different soak time were compared in table 2. When the device was in air, the frequency at powered-up was at 910.5 MHz with RF signal strength of -80 dBm measured using a spectrum analyzer. The RF signal strength increased to -75 dBm after the immersion of the device in PBS solution (table 2). The custom-built hand receiver confirmed the increase of RF strength from -61 to -47 dBm after submerging the device into PBS. The initial increase in RF signal strength is most likely due to the change of media from air to PBS solution, and has been observed previously. The discrepancies between the two RF signal strengths measured by spectrum analyzer and hand receiver were expected due to the differences in antennas and electronics.

The long-term RF signal strengths and their corresponding frequencies are presented in figure 4 as a function of soak time. The power-up frequency was continuously near 910 MHz and the RF signal strength was stably around -73 dBm (figure 4 (b)) during the equivalent soaking time of 1044 days at 37 °C. The small fluctuations in RF signal strengths and respective frequencies could be caused by environmental noise and the different positions and distances between the reference wire and antenna. This represents a considerably longer soak test results compared with Sharma *et al.*'s reporting of a lifetime of 276 days (lasted ~ 500 days with unpolished data) at room temperature with Parylene encapsulation [18]. The room temperature soak testing could be considered as a "decelerated" lifetime testing with aging factor of 0.35, which gave an equivalent lifetime of 100 days at 37 °C. The bilayer coated devices are still under soak testing. The long-term stability of power-up frequencies and RF signal suggests the good insulation of the alumina and Parylene C bilayer encapsulation for implantable devices.

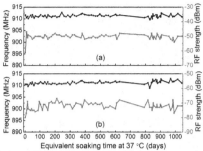

Equivalent soaking time at 37 °C (days)

Figure 4. Transmitted wireless RF signal strength and frequency monitored as a function of soak time in PBS. (a) Peak RF signal strengths and the respective frequencies as extracted from the spectra measured using a customized wireless hand receiver unit. (b) RF signal strengths and the respective frequencies as monitored from a spectrum analyzer.

Table 2. Radio-frequency (RF) signal strengths and frequencies of the wireless INIR-6 device measured in PBS using a customized wireless hand receiver unit and a spectrum analyzer.

Soak time	RF from spectrum analyzer		RF from hand receiver	
	Frequency (MHz)	Signal Strength (dBm)	Frequency (MHz)	Signal Strength (dBm)
0 (in air)	910.5	-80	911.6	-61
1 day	910.5	-75	910.5	-47
300 days	910.3	-71	910.7	-51
1044 days	911	-72	910.8	-50

CONCLUSIONS

In summary, we have demonstrated the long-term reliability and stability of ALD alumina and Parylene C bilayer coated neural interfaces. Wired and wireless UEAs were soaked in PBS at 57 °C for accelerated lifetime testing. Different from the trend of continuous drop in impedance for Parylene-only coated arrays, median impedances of alumina and Parylene bilayer coated wired arrays increased from 60 kΩ to 160 kΩ after 960 equivalent days of soak testing at 37 °C. Bilayer coated wireless UEAs incorporated with active electronics had stable power-up frequency of ~ 910 MHz and constant RF signal strength of ~ -50 dBm (measured by hand receiver) over 1044 equivalent days of soak testing at 37 °C, showing the slow water ingress and excellent insulation performance of the bilayer encapsulation. Based on the coating performance on neural interfaces, it is believed that this bilayer encapsulation can be used for many other chronic biomedical implantable devices to improve the lifetime of those devices.

ACKNOWLEDGMENTS

The authors would like to thank Fraunhofer IZM for the integration of the wireless devices. Funding of this research is provided by DARPA contract No: N66001-06-C-4056 and NIH contract No: 1R01NS064318-01A1. Sandeep Negi and Florian Solzbacher have financial interest in Blackrock microsystems. The views expressed are those of the authors and do not reflect the official policy or position of the Department of Defense or the U.S. Government. Approved for public release; distribution unlimited.

REFERENCES

[1] R. A. Normann, "Technology Insight: Future neuroprosthetic therapies for disorders of the nervous system," *Nature Clinical Practice Neurology*, vol. 3, pp. 444-452, 2007.

[2] X. Z. Xie, L. Rieth, P. Tathireddy, and F. Solzbacher, "Long-term in-vivo Investigation of Parylene-C as Encapsulation Material for Neural Interfaces," *Procedia Engineering*, vol. 25, pp. 483-486, 2011.

[3] M. Guenther, G. Gerlach, T. Wallmersperger, M. N. Avula, S. H. Cho, X. Xie, *et al.*, "Smart Hydrogel-Based Biochemical Microsensor Array for Medical Diagnostics," *Advances in Science and Technology*, vol. 85, pp. 47-52, 2013.

[4] W. Li, D. C. Rodger, P. Menon, and Y. C. Tai, "Corrosion Behavior of Parylene-Metal-Parylene Thin Films in Saline," *ECS Transactions*, vol. 11, pp. 1-6, 2008.

[5] E. Langereis, M. Creatore, S. Heil, M. Van de Sanden, and W. Kessels, "Plasma-assisted atomic layer deposition of Al_2O_3 moisture permeation barriers on polymers," *Applied physics letters*, vol. 89, pp. 081915-081915-3, 2006.

[6] X. Xie, L. Rieth, S. Merugu, P. Tathireddy, and F. Solzbacher, "Plasma-assisted atomic layer deposition of Al2O3 and parylene C bi-layer encapsulation for chronic implantable electronics," *Applied physics letters*, vol. 101, pp. 093702 1-5, 2012.

[7] X. Xie, L. Rieth, R. Caldwell, M. Diwekar, P. Tathireddy, R. Sharma, and F. Solzbacher, "Long-Term Bilayer Encapsulation Performance of Atomic Layer Deposited Al2O3 and Parylene C for Biomedical Implantable Devices," *Biomedical Engineering, IEEE Transactions on*, vol. 60, pp. 2943-2951, 2013.

[8] S. Minnikanti, G. Diao, J. J. Pancrazio, X. Xie, L. Rieth, F. Solzbacher, and N. Peixoto, "Lifetime assessment of atomic-layer-deposited Al2O3–Parylene C bilayer coating for neural interfaces using accelerated age testing and electrochemical characterization," *Acta Biomaterialia*, vol. 10, pp. 960-967, 2014.

[9] X. Xie, L. Rieth, P. Tathireddy, and F. Solzbacher, "Atomic layer deposited Al2O3 and parylene C dual-layer encapsulation for biomedical implantable devices," in *Solid-State Sensors, Actuators and Microsystems (TRANSDUCERS & EUROSENSORS XXVII), 2013 Transducers & Eurosensors XXVII: The 17th International Conference on*, 2013, pp. 1044-1047.

[10] R. Bhandari, S. Negi, L. Rieth, and F. Solzbacher, "A wafer-scale etching technique for high aspect ratio implantable MEMS structures," *Sensors and Actuators, A: Physical*, vol. 162, pp. 130-136, 2010.

[11] S. Kim, R. Bhandari, M. Klein, S. Negi, L. Rieth, P. Tathireddy, M. Toepper, H. Oppermann, and F. Solzbacher, "Integrated wireless neural interface based on the Utah electrode array," *Biomedical microdevices*, vol. 11, pp. 453-466, 2009.

[12] X. Xie, L. Rieth, S. Negi, R. Bhandari, R. Caldwell, R. Sharma, P. Tathireddy, and F. Solzbacher, "Self-aligned tip deinsulation of atomic layer deposited Al 2 O 3 and parylene

C coated Utah electrode array based neural interfaces," *Journal of Micromechanics and Microengineering,* vol. 24, p. 035003, 2014.

[13] X. Xie, L. Rieth, R. Cardwell, R. Sharma, J. M. Yoo, M. Diweka, P. Tathireddy, and F. Solzbacher, "Bi-layer encapsulation of utah array based nerual interfaces by atomic layer deposited Al2O3 and parylene C," in *Solid-State Sensors, Actuators and Microsystems (TRANSDUCERS & EUROSENSORS XXVII), 2013 Transducers & Eurosensors XXVII: The 17th International Conference on,* 2013, pp. 1267-1270.

[14] K. Hemmerich, "General aging theory and simplified protocol for accelerated aging of medical devices," *MEDICAL PLASTIC AND BIOMATERIALS,* vol. 5, pp. 16-23, 1998.

[15] R. R. Harrison, R. J. Kier, C. A. Chestek, V. Gilja, P. Nuyujukian, S. Ryu, B. Greger, F. Solzbacher, and K. V. Shenoy, "Wireless neural recording with single low-power integrated circuit," *IEEE Transactions on Neural Systems and Rehabilitation Engineering,* vol. 17, pp. 322-329, 2009.

[16] J. M. Hsu, L. Rieth, R. A. Normann, P. Tathireddy, and F. Solzbacher, "Encapsulation of an integrated neural interface device with Parylene C," *Biomedical Engineering, IEEE Transactions on,* vol. 56, pp. 23-29, 2009.

[17] S. Kane, S. Cogan, J. Ehrlich, T. Plante, and D. McCreery, "Electrical performance of penetrating microelectrodes chronically implanted in cat cortex," in *Engineering in Medicine and Biology Society, EMBC, 2011 Annual International Conference of the IEEE,* 2011, pp. 5416-5419.

[18] A. Sharma, L. Rieth, P. Tathireddy, R. Harrison, H. Oppermann, M. Klein, *et al.,* "Long term in vitro functional stability and recording longevity of fully integrated wireless neural interfaces based on the Utah Slant Electrode Array," *Journal of Neural Engineering,* vol. 8, 2011.

AUTHOR INDEX

SUBJECT INDEX

Printed in the United States
by Baker & Taylor Publisher Services